Editorial project:
© 2023 **booq** publishing, S.L.
c/ Domènech, 7-9, 2º 1ª
08012 Barcelona, Spain
T: +34 93 268 80 88
www.booqpublishing.com

ISBN 978-84-9936-646-3 [EN]
ISBN 978-84-9936-636-4 [ES]

Editorial coordinator:
Claudia Martínez Alonso

Art director:
Mireia Casanovas Soley

Editor:
Sergio Asensio

Layout:
Cristina Simó Perales

Translation:
© **booq** publishing, S.L.

Printing in Spain

Down to Earth: Rammed Earth Architecture explores the beauty and versatility of rammed earth construction technique, also known as rammed earth. Through a compilation of 22 studies from around the world, this book presents some of the best projects in rammed earth architecture, ranging from modern designs to traditional buildings, all with a focus on sustainability.

Rammed earth construction is an ancient technique that has been used worldwide for centuries. Nowadays, this technique has been revitalized as a sustainable and low-cost building solution for the 21st century. In this book, architects present their projects and explain how they have used this technique to create beautiful and sustainable buildings that blend harmoniously into the natural environment.

Down to Earth: Rammed Earth Architecture features a wide variety of rammed earth architecture projects, from houses that seamlessly integrate with the nature of GLH Architects in South Africa, to an impressive primary care center in Australia by Kaunitz Yeung Architecture. Each of these projects demonstrates the ability of rammed earth to create impressive and durable buildings that blend harmoniously into their natural environment.

Down to Earth: Rammed Earth Architecture erforscht die Schönheit und Vielseitigkeit der Bauweise in verdichteter Erde, auch bekannt als rammed earth. Durch eine Zusammenstellung von 22 Studien aus der ganzen Welt präsentiert dieses Buch einige der besten Architekturprojekte in verdichteter Erde, von modernen Designs bis hin zu traditionellen Gebäuden, alle mit Fokus auf Nachhaltigkeit.

Die Bauweise in verdichteter Erde ist eine alte Technik, die seit Jahrhunderten weltweit genutzt wird. Heute wird diese Technik als nachhaltige und kostengünstige Baulösung für das 21. Jahrhundert wiederbelebt. In diesem Buch präsentieren Architekten ihre Projekte und erklären, wie sie diese Technik verwendet haben, um schöne und nachhaltige Gebäude zu schaffen, die harmonisch in die natürliche Umgebung integriert sind.

Down to Earth: Rammed Earth Architecture präsentiert eine Vielzahl von Architekturprojekten in verdichteter Erde, von Häusern, die perfekt in die Natur von Südafrika von GLH Architects integriert sind, bis hin zu einem beeindruckenden Primärversorgungszentrum in Australien von Kaunitz Yeung Architecture. Jedes dieser Projekte zeigt die Fähigkeit der verdichteten Erde, beeindruckende und langlebige Gebäude zu schaffen, die harmonisch in ihre natürliche Umgebung integriert sind.

Down to Earth: Rammed Earth Architecture explore la beauté et la polyvalence de la technique de construction en terre battue, également connue sous le nom de rammed earth. À travers une compilation de 22 études du monde entier, ce livre présente certains des meilleurs projets d'architecture en terre battue, allant de designs modernes à des bâtiments traditionnels, tous axés sur la durabilité.

La construction en terre battue est une technique ancienne utilisée dans le monde entier depuis des siècles. Aujourd'hui, cette technique a été revitalisée en tant que solution de construction durable et peu coûteuse pour le XXIe siècle. Dans ce livre, les architectes présentent leurs projets et expliquent comment ils ont utilisé cette technique pour créer des bâtiments beaux et durables qui s'intègrent harmonieusement dans l'environnement naturel.

Down to Earth: Rammed Earth Architecture présente une large variété de projets d'architecture en terre battue, allant de maisons qui s'intègrent parfaitement dans la nature d'Afrique du Sud de GLH Architects à un impressionnant centre de soins primaires en Australie de Kaunitz Yeung Architecture. Chacun de ces projets démontre la capacité de la terre battue à créer des bâtiments impressionnants et durables qui s'intègrent harmonieusement dans leur environnement naturel.

Tierra: arquitectura natural explora la belleza y la versatilidad de la técnica de construcción en tierra apisonada, también conocida como rammed earth. A través de una recopilación de 22 estudios de todo el mundo, este libro presenta algunos de los mejores proyectos de arquitectura en tierra apisonada, desde diseños modernos hasta edificaciones tradicionales, y todos ellos con un enfoque en la sostenibilidad.

La construcción en tierra apisonada es una técnica antigua que ha sido utilizada en todo el mundo durante siglos. Hoy en día, esta técnica se ha revitalizado como una solución de construcción sostenible y de bajo costo para el siglo XXI. En este libro, los arquitectos presentan sus proyectos y explican cómo han utilizado esta técnica para crear edificaciones hermosas y sostenibles que se integran armoniosamente en el entorno natural.

Tierra: arquitectura natural presenta una amplia variedad de proyectos de arquitectura en tierra apisonada, que van desde casas que se integran perfectamente en la naturaleza de Sudáfrica de GLH Architects, hasta un impresionante centro de atención primaria en Australia de Kaunitz Yeung Architecture. Cada uno de estos proyectos demuestra la capacidad de la tierra apisonada para crear edificaciones impresionantes y duraderas que se integran armoniosamente en su entorno natural.

AAE | ARQUITECTURA // ARTURO AGUILAR ESPINOZA - JAVIER DÍAZ CERÓN

AAE is an Architecture Office based in Mexico City and headed by Arturo Aguilar Espinoza and Javier Díaz Cerón, who have developed high quality architectural projects in the residential, commercial and interior design sectors, exposing a contemporary language through the use of natural and authentic materials. With the main objective of innovation in construction systems and the use of local resources, the studio projects a vision of timeless architecture, seeking a balance between form and function.

AAE ist ein Architekturbüro mit Sitz in Mexiko-Stadt, das von Arturo Aguilar Espinoza und Javier Díaz Cerón geleitet wird. Sie haben hochwertige Architekturprojekte in den Bereichen Wohnen, Gewerbe und Innenarchitektur entwickelt, die durch die Verwendung natürlicher und authentischer Materialien eine zeitgemäße Sprache sprechen. Mit dem Hauptziel der Innovation von Bausystemen und der Nutzung lokaler Ressourcen entwirft das Studio eine Vision zeitloser Architektur, die ein Gleichgewicht zwischen Form und Funktion anstrebt.

AAE est un bureau d'architecture basé à Mexico et dirigé par Arturo Aguilar Espinoza et Javier Díaz Cerón, qui ont développé des projets architecturaux de haute qualité dans les secteurs résidentiel, commercial et de l'architecture d'intérieur, exposant un langage contemporain à travers l'utilisation de matériaux naturels et authentiques. Avec pour objectif principal l'innovation dans les systèmes de construction et l'utilisation des ressources locales, le studio projette une vision de l'architecture intemporelle, à la recherche d'un équilibre entre la forme et la fonction.

AAE es una Oficina de Arquitectura con sede en la Ciudad de México y encabezada por Arturo Aguilar Espinoza y Javier Díaz Cerón, que han desarrollado proyectos de alta calidad arquitectónica en el sector residencial, comercial e interiorismo, exponiendo un lenguaje contemporáneo mediante el uso de materiales naturales y auténticos. Teniendo como principal objetivo la innovación en sistemas constructivos y el aprovechamiento de recursos locales, el estudio proyecta una visión de arquitectura atemporal, buscando el equilibrio entre forma y función.

CASA TIERRA

CIUDAD DE MÉXICO, MEXICO // PHOTOS © JAVIER DÍAZ

Design Architects: Arturo Aguilar Espinoza, Javier Díaz Cerón
Client: Private
Client Project Managers: AAE arquitectura
Centre: Bosque Real, México
The Collaborative Artwork by: Raíz Arquitectura, Karen Poulain, Gabriela Amicone
Landscape Architect: AAE arquitectura
Civil Engineer: Juan Marcos Hernández Horroa

This house uses a biophilic concept that focuses on the use of natural materials, especially earth, to build hierarchical walls both inside and outside the house. Circulations are designed to highlight architectural elements and are accompanied by natural ones. From the rooftop, views of the city are enjoyed, and social spaces are created with gardens and orchards. The house has a double-height study-library, as well as intimate spaces with views of garden areas or city landscapes. Casa Tierra is part of a pair of houses that share a central garden for coexistence, but maintain the privacy of each one. Located in an exclusive subdivision in Mexico City, this house is a great investment to live in an architectural and natural space that promotes the use of natural materials with a low carbon footprint and timeless design.

Dieses Haus nutzt ein biofilisches Konzept, das sich auf den Einsatz natürlicher Materialien konzentriert, insbesondere Erde, um hierarchische Wände im Innen- und Außenbereich des Hauses zu bauen. Die Zirkulationen werden entworfen, um die architektonischen Elemente hervorzuheben und werden von natürlichen Elementen begleitet. Von der Dachterrasse aus können Sie Aussicht auf die Stadt genießen und es werden soziale Räume mit Gärten und Gemüsegärten geschaffen. Das Haus verfügt über ein zweigeschossiges Studio-Bibliothek sowie intime Räume mit Blick auf Gartenbereiche oder Stadtbilder. Das Casa Tierra ist Teil eines Paares von Häusern, die einen zentralen Garten für das Zusammenleben teilen, aber die Privatsphäre jeder bewahren. In einer exklusiven Wohngegend von Mexiko-Stadt gelegen, ist dieses Haus eine großartige Investition für ein architektonisches und natürliches Umfeld, das den Einsatz natürlicher Materialien mit geringer Kohlenstoffbilanz und zeitlosem Design fördert.

Cette maison utilise un concept biophilique qui se concentre sur l'utilisation de matériaux naturels, en particulier de la terre, pour construire des murs hiérarchiques à l'intérieur et à l'extérieur de la maison. Les circulations sont conçues pour mettre en valeur les éléments architecturaux et sont accompagnées d'éléments naturels. Depuis le toit, les vues sur la ville sont appréciées et des espaces sociaux sont créés avec des jardins et des vergers. La maison dispose d'une bibliothèque-étude à double hauteur, ainsi que d'espaces intimes avec vue sur les zones de jardin ou les paysages urbains. Casa Tierra fait partie d'une paire de maisons qui partagent un jardin central pour la coexistence, mais qui maintiennent la vie privée de chacune. Située dans une subdivision exclusive de Mexico, cette maison est un excellent investissement pour vivre dans un espace architectural et naturel qui favorise l'utilisation de matériaux naturels à faible empreinte carbone et à design intemporel.

Esta casa utiliza un concepto biofílico que se enfoca en el uso de materiales naturales, especialmente la tierra, para construir muros jerárquicos en el interior y exterior de la casa. Las circulaciones se diseñan para resaltar los elementos arquitectónicos y están acompañadas por elementos naturales. Desde la azotea, se puede disfrutar de vistas hacia la ciudad y se crean espacios sociales con jardines y huertos. La casa tiene un estudio-biblioteca de doble altura, así como espacios íntimos con vistas hacia áreas ajardinadas o paisajes de la ciudad. La Casa Tierra forma parte de un par de viviendas que comparten un jardín central para la convivencia, pero mantienen la privacidad de cada una. Ubicada en un fraccionamiento exclusivo de la Ciudad de México, esta casa es una gran inversión para vivir en un espacio arquitectónico y natural, que promueve el uso de materiales naturales con baja huella de carbono y con un diseño atemporal.

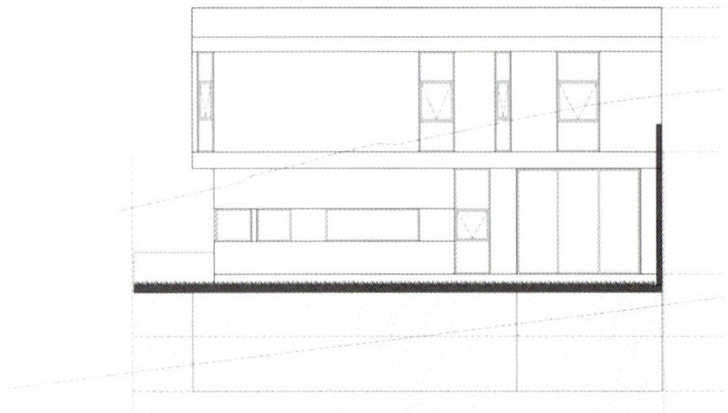

North elevation

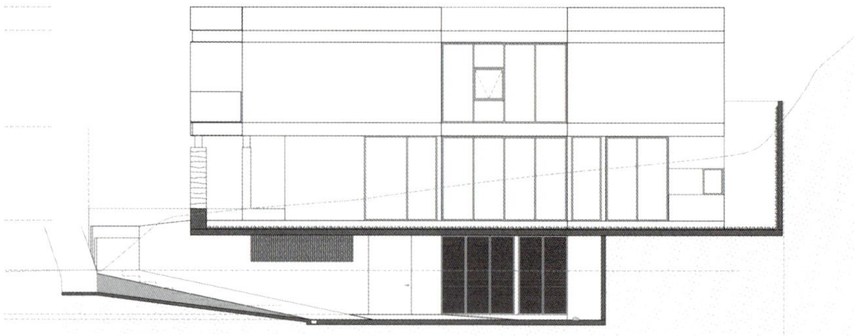

East elevation

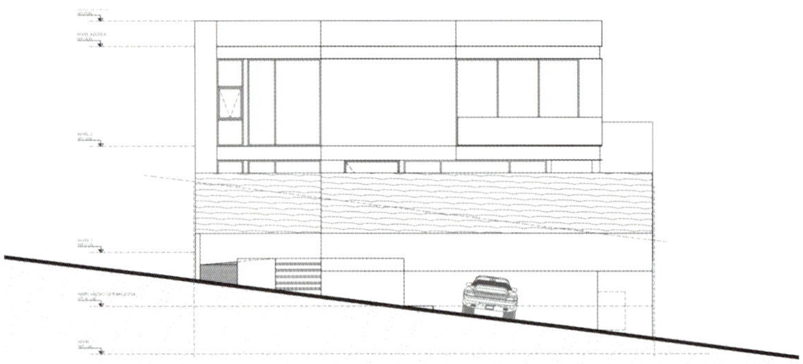

South elevation

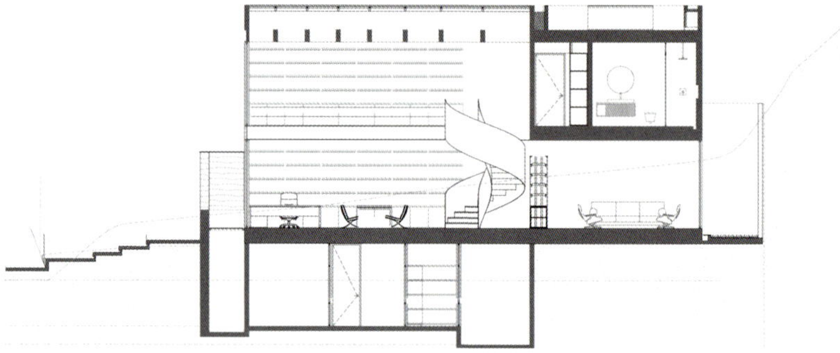

A-A' section

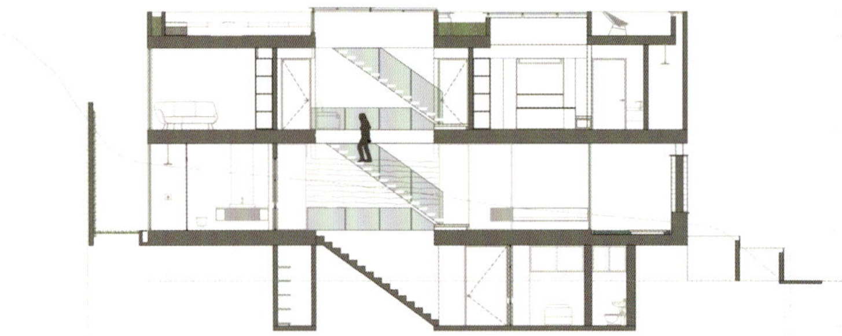

B-B' section

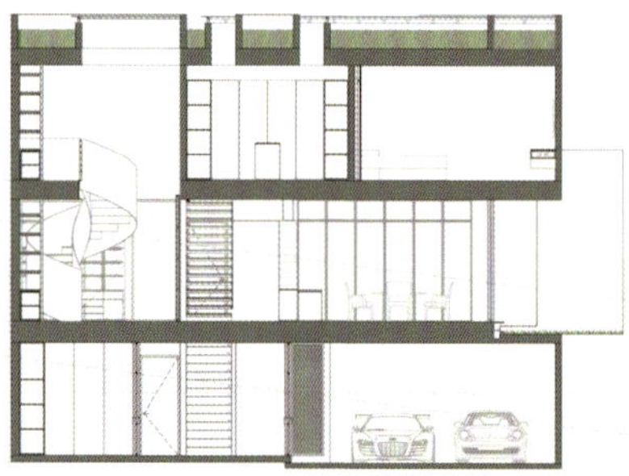

C-C' section

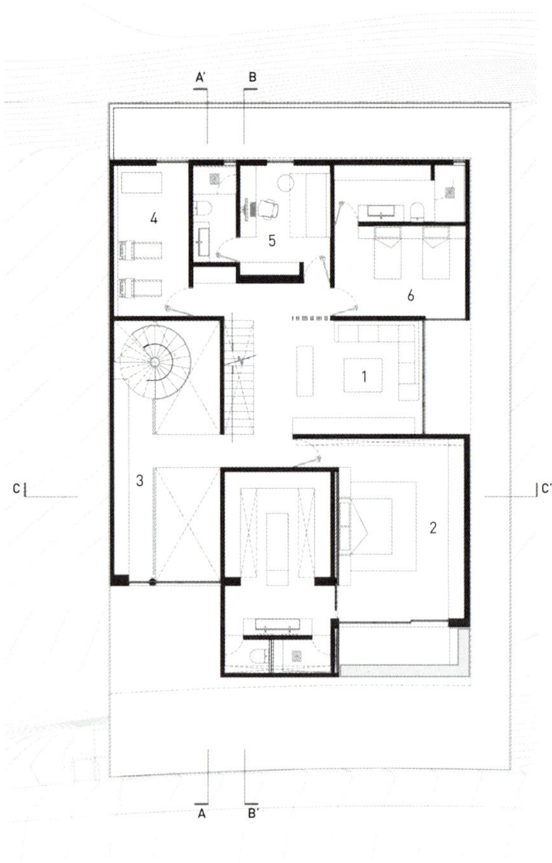

1. Family room
2. Master bedroom
3. Library
4. Gym
5. Study
6. Bedroom

Second level plan

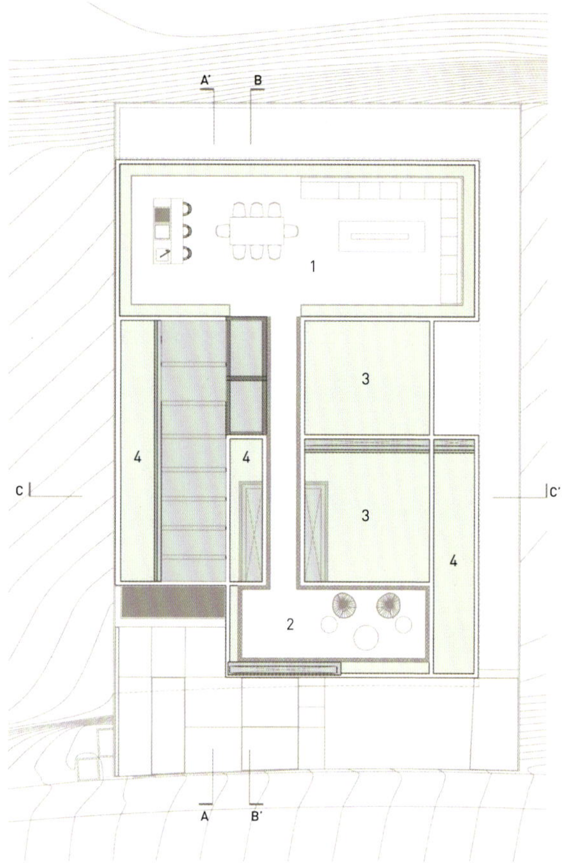

1. Terrace
2. Lookout
3. Orchard
4. Garden

Rooftop plan

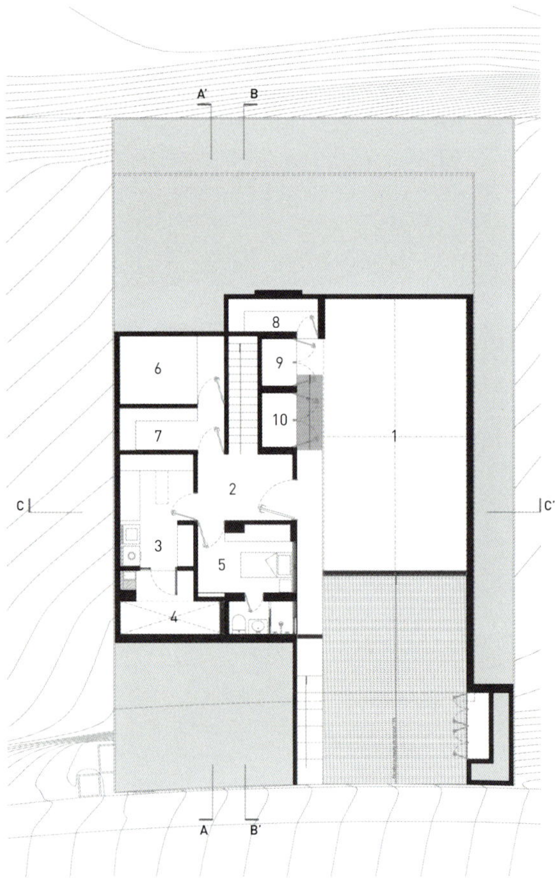

1. Garage
2. Hall
3. Laundry
4. Patio
5. Maiden room
6. Mechanical room
7. Cellar
8. Store
9. Boilers
10. Trash

Basement plan

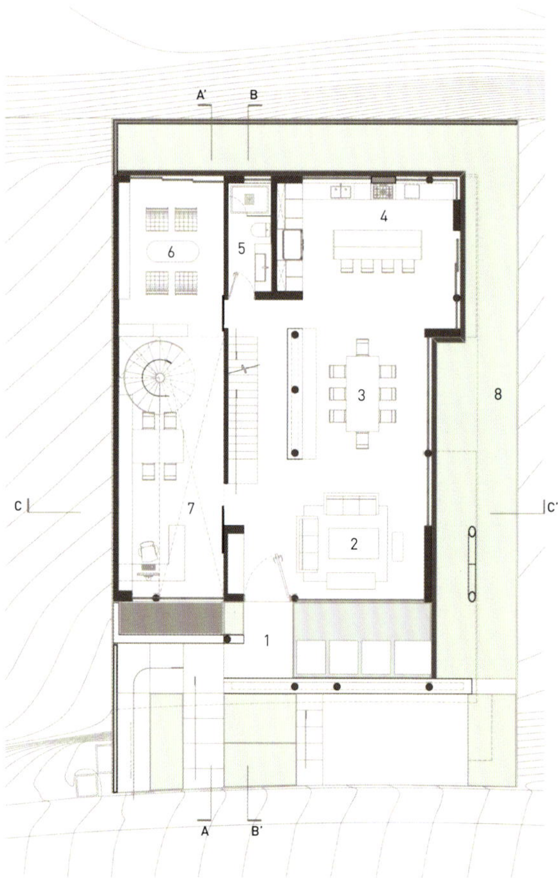

1. Hall
2. Living room
3. Dining room
4. Kitchen
5. Guest bathroom
6. chess room
7. Office
8. Garden

First level plan

AIDLIN DARLING DESIGN // DAVID DARLING - JOSHUA AIDLIN

From a shared interest in exploring design from a wide range of scales, programs and disciplines, partners Joshua Aidlin and David Darling started Aidlin Darling Design around a woodworking shop in 1997. The award-winning studio acts as a creative hub for a network of collaborators, including artists, builders, craftspeople, engineers and chefs, with a firm belief that design can enlighten the human spirit by engaging all the senses.
Since its origination, Aidlin Darling Design has collected more than 200 regional, national and international awards, including the inaugural AIA/COTE Top Ten Plus Green Project Award, two international Civic Trust Awards, seven national AIA Awards and two James Beard Awards. In 2013, the firm received the Smithsonian's Cooper-Hewitt Museum's National Design Award for all of its work.

Die Partner Joshua Aidlin und David Darling gründeten Aidlin Darling Design 1997 in einer Holzwerkstatt aus dem gemeinsamen Interesse heraus, Design in verschiedenen Maßstäben, Programmen und Disziplinen zu erforschen. Das preisgekrönte Studio fungiert als kreativer Knotenpunkt für ein Netzwerk von Mitarbeitern, darunter Künstler, Bauunternehmer, Handwerker, Ingenieure und Köche, in der festen Überzeugung, dass Design den menschlichen Geist erleuchten kann, indem es alle Sinne anspricht.
Seit seiner Gründung hat Aidlin Darling Design mehr als 200 regionale, nationale und internationale Auszeichnungen erhalten, darunter den ersten AIA/COTE Top Ten Plus Green Project Award, zwei internationale Civic Trust Awards, sieben nationale AIA Awards und zwei James Beard Awards. Im Jahr 2013 erhielt das Unternehmen den National Design Award des Smithsonian Cooper-Hewitt Museum für alle seine Arbeiten.

HOUSE AND EARTH & SKY

Partant d'un intérêt commun pour l'exploration du design à partir d'un large éventail d'échelles, de programmes et de disciplines, les partenaires Joshua Aidlin et David Darling ont créé Aidlin Darling Design autour d'un atelier de menuiserie en 1997. Le studio primé fait office de centre de création pour un réseau de collaborateurs, dont des artistes, des constructeurs, des artisans, des ingénieurs et des chefs, avec la ferme conviction que le design peut éclairer l'esprit humain en faisant appel à tous les sens.

Depuis sa création, Aidlin Darling Design a récolté plus de 200 prix régionaux, nationaux et internationaux, dont le premier prix AIA/COTE Top Ten Plus Green Project, deux prix internationaux Civic Trust, sept prix nationaux AIA et deux prix James Beard. En 2013, le cabinet a reçu le National Design Award du Smithsonian's Cooper-Hewitt Museum pour l'ensemble de son travail.

A partir de un interés compartido por explorar el diseño desde una amplia gama de escalas, programas y disciplinas, los socios Joshua Aidlin y David Darling iniciaron Aidlin Darling Design en torno a un taller de carpintería en 1997. El galardonado estudio actúa como un centro creativo para una red de colaboradores, incluyendo artistas, constructores, artesanos, ingenieros y chefs, con la firme convicción de que el diseño puede iluminar el espíritu humano al involucrar todos los sentidos.

Desde su creación, Aidlin Darling Design ha recogido más de 200 premios regionales, nacionales e internacionales, entre ellos el primer premio AIA/COTE Top Ten Plus Green Project, dos premios internacionales Civic Trust, siete premios nacionales AIA y dos premios James Beard. En 2013, la firma recibió el Premio Nacional de Diseño del Museo Cooper-Hewitt del Smithsonian por toda su obra.

WINDHOVER CONTEMPLATIVE CENTER

HOUSE AND EARTH & SKY

HILLSBOROUGH, CALIFORNIA, UNITED STATES
PHOTOS © MATTHEW MILLMAN PHOTOGRAPHY

AWARDS:

- Luxe Magazine RED Awards
  2018 Award Winner for Contemporary Design
- Society of American Registered Architects
  2017 Excellence Award
- AIA California
  2017 Architecture Merit Award
- The Chicago Athenaeum Museum
  of Architecture & Design
  2017 Green Good Design Award
- AIA California
  2016 Residential Design Merit Award

Completion: June 2015
Size: 7,477 gross square feet,
5,609 conditioned square feet
Clients: Bay Chang & Sally Liu
General Contractor: Ryan Associates
Landscape Architect: Lutsko Associates
Structural Engineer: GFDS Engineers
Lighting Design: EJA Lighting Design
Geotechnical Engineer: Rollo & Ridley
Civil Engineer: Lea & Braze Engineering
Interior Design: Gary Hutton Design
Mechanical Engineer: Integral Group
Rammed Earth Consultant: Rammed Earth Works
LEED Consultant: Gilleran Energy Management
Specifications: White & Green Specs
Rainwater & Greywater Consultant: WaterSprout

This sustainable single-family home uses rammed earth walls and kite-like floating roofs to create indoor-outdoor living spaces that celebrate the intersection between earth and sky. The extensive use of glass blurs the boundary between inside and outside, creating a natural garden-like environment. The owners set ambitious sustainability goals, striving for LEED Platinum certification and Net Zero Energy, and the design integrates ecological responsibility holistically. The rammed earth walls provide thermal mass and bracket space, while the folded roof planes integrate multiple sustainable design strategies. The home is designed for Net Zero Energy and uses passive systems to address light and temperature, with electric-only mechanical systems providing balance. The goal is to create a sustainable building that endures, is beautiful and comfortable, and is a delight to live in for generations to come.

Dieses nachhaltige Einfamilienhaus nutzt stampflehmwände und drachenähnliche schwimmende Dächer, um Innen-Außen-Lebensräume zu schaffen, die das Zusammentreffen von Erde und Himmel feiern. Der umfangreiche Einsatz von Glas verwischt die Grenze zwischen Innen und Außen und schafft eine natürliche Garten-Umgebung. Die Besitzer haben ehrgeizige Nachhaltigkeitsziele festgelegt und streben nach LEED Platinum-Zertifizierung und Netto-Nullenergie, wobei das Design die ökologische Verantwortung ganzheitlich integriert. Die stampflehmwände bieten thermische Masse und gliedern den Raum, während die gefalteten Dachflächen mehrere nachhaltige Designstrategien integrieren. Das Haus ist für Netto-Nullenergie konzipiert und verwendet passive Systeme zur Steuerung von Licht und Temperatur, wobei elektrisch betriebene mechanische Systeme das Gleichgewicht herstellen. Das Ziel ist es, ein nachhaltiges Gebäude zu schaffen, das beständig ist, schön und komfortabel ist und über lange Zeit Freude bereitet.

Cette maison unifamiliale durable utilise des murs en terre compactée et des toits flottants en forme de cerf-volant pour créer des espaces de vie intérieurs-extérieurs qui célèbrent l'intersection entre la terre et le ciel. L'utilisation intensive de verre brouille la frontière entre l'intérieur et l'extérieur, créant un environnement semblable à un jardin naturel. Les propriétaires ont fixé des objectifs ambitieux de durabilité, luttant pour la certification LEED Platinum et Net Zero Energy, et la conception intègre la responsabilité écologique de manière holistique. Les murs en terre compactée fournissent de la masse thermique et délimitent l'espace, tandis que les plans de toit pliés intègrent de multiples stratégies de conception durables. La maison est conçue pour une énergie nette zéro et utilise des systèmes passifs pour aborder la lumière et la température, avec des systèmes mécaniques uniquement électriques fournissant un équilibre. Le but est de créer un bâtiment durable qui perdure, est beau et confortable, et est un plaisir de vivre pour les générations à venir.

Esta casa unifamiliar sostenible utiliza muros de tierra compactada y techos flotantes en forma de cometa para crear espacios de vida interior-exterior que celebran la intersección entre la tierra y el cielo. El uso extensivo de vidrio borra la frontera entre el interior y el exterior, creando un ambiente similar a un jardín natural. Los propietarios establecieron objetivos ambiciosos de sostenibilidad, luchando por la certificación LEED Platinum y Net Zero Energy, y el diseño integra la responsabilidad ecológica de manera global. Los muros de tierra compactada proporcionan masa térmica y delimitan el espacio, mientras que los techos plegados integran múltiples estrategias de diseño sostenible. La casa está diseñada para Net Zero Energy y utiliza sistemas pasivos para abordar la luz y la temperatura, con sistemas mecánicos solo eléctricos que proporcionan equilibrio. El objetivo es crear un edificio sostenible que perdure, sea hermoso y cómodo, y que sea un placer vivir en él durante generaciones.

Render section

Sketch

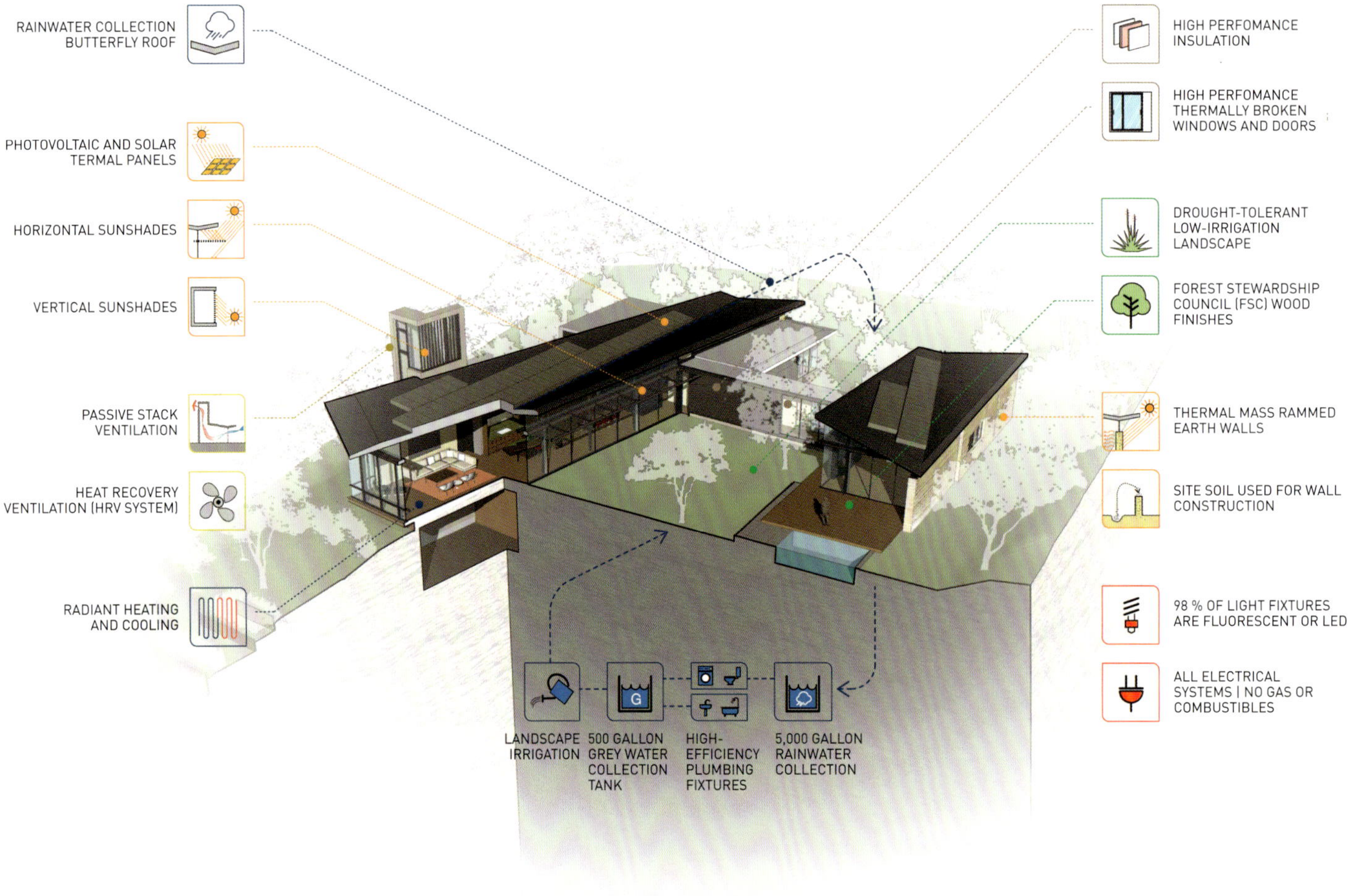

Sustainability diagram

WINDHOVER CONTEMPLATIVE CENTER

STANFORD, CALIFORNIA, UNITED STATES
PHOTOS © MATTHEW MILLMAN PHOTOGRAPHY

AWARDS:

- The Chicago Athenaeum Museum of Architecture & Design 2015 American Architecture Award
- AIA National 2016 Interior Architecture Award
- Civic Trust Awards 2016 Winner
- IIDA National 2015 Interior Design Competition, Best of Competition Winner
- ASLA National 2017 Honor Award

Client: Stanford University
Campus Architect: David Lenox
Campus Preservation Architect: Sapna Marfatia
Stanford Project Manager: Maggie Burgett
Landscape Architect: Andrea Cochran Landscape Architecture
Rammed Earth Fabricator: Rammed Earth Works
Structural Engineer: Rutherford + Chekene
Lighting Design: Auerbach Glasow
Daylighting Consultant: Loisos + Ubbelohde
Acoustical Consultant: Salter
General Contractor: SC Builders

The Windhover Contemplative Center is a spiritual retreat on the Stanford campus, designed to offer a refuge for personal renewal and quiet reflection to Stanford students, faculty, staff, and the wider community. The center is inspired by Nathan Oliveira's Windhover paintings, which were created after the artist saw kestrels flying over the Stanford foothills. The center unifies art, landscape, and architecture to replenish and invigorate the spirit. The building is located in the heart of the campus, adjacent to a natural oak grove, and is surrounded by a private garden lined with tall bamboo. The space is intentionally kept dark, with louvered skylights providing natural light to highlight the paintings and the landscape beyond. Thick rammed earth walls and wood surfaces heighten the sensory experience, while water features aid contemplation. The center is accessible day and night, and visitors can even view the paintings from the oak grove without entering the building.

Das Windhover Contemplative Center ist ein spirituelles Rückzugszentrum auf dem Stanford-Campus, das entworfen wurde, um Stanford-Studenten, Fakultätsmitgliedern, Mitarbeitern und der breiteren Gemeinschaft einen Rückzugsort für persönliche Erneuerung und ruhige Besinnung zu bieten. Das Zentrum ist von Nathan Oliveiras Windhover-Gemälden inspiriert, die entstanden, als der Künstler Turmfalken über den Stanford-Hügeln sah. Das Zentrum vereint Kunst, Landschaft und Architektur, um den Geist zu erneuern und zu beleben. Das Gebäude befindet sich im Herzen des Campus, neben einem natürlichen Eichenhain, und ist von einem privaten Garten umgeben, der von hohen Bambuspflanzen gesäumt ist. Der Raum ist absichtlich dunkel gehalten, wobei Lamellen-Skylights Tageslicht bieten, um die Gemälde und die Landschaft zu betonen. Dicke Lehmwände und Holzoberflächen erhöhen das sensorische Erlebnis, während Wassermerkmale zur Besinnung beitragen. Das Zentrum ist tagsüber und nachts zugänglich, und Besucher können sogar die Gemälde vom Eichenhain aus betrachten, ohne das Gebäude zu betreten.

Le Centre de Contemplation Windhover est une retraite spirituelle sur le campus de Stanford, conçue pour offrir un refuge de renouveau personnel et de réflexion tranquille aux étudiants, enseignants, personnel et à la communauté en général. Le centre est inspiré des peintures de Windhover de Nathan Oliveira, créées après que l'artiste a vu des crécerelles voler au-dessus des collines de Stanford. Le centre unifie l'art, le paysage et l'architecture pour reconstituer et revigorer l'esprit. Le bâtiment est situé au cœur du campus, à côté d'une forêt naturelle de chênes, et est entouré d'un jardin privé bordé de bambous hauts. L'espace est intentionnellement sombre, avec des puits de lumière à lames qui fournissent une lumière naturelle pour mettre en valeur les peintures et le paysage au-delà. Les murs épais en terre battue et les surfaces en bois intensifient l'expérience sensorielle, tandis que les éléments d'eau aident à la contemplation. Le centre est accessible jour et nuit, et les visiteurs peuvent même voir les peintures depuis la forêt de chênes sans entrer dans le bâtiment.

El Centro de Contemplación Windhover es un refugio espiritual en el campus de Stanford, diseñado para ofrecer un refugio de renovación personal y reflexión tranquila para estudiantes, profesores, personal y la comunidad en general. El centro está inspirado en las pinturas de Windhover de Nathan Oliveira, que fueron creadas después de que el artista viera a los cernícalos volando sobre las colinas de Stanford. El centro unifica el arte, el paisaje y la arquitectura para reponer e invigorar el espíritu. El edificio está ubicado en el corazón del campus, junto a un bosque natural de robles, y está rodeado por un jardín privado revestido con bambú alto. El espacio se mantiene intencionalmente oscuro, con tragaluces con lamas que proporcionan luz natural para resaltar las pinturas y el paisaje más allá. Las gruesas paredes de tierra compactada y las superficies de madera aumentan la experiencia sensorial, mientras que las características del agua ayudan a la contemplación. El centro es accesible de día y de noche, y los visitantes incluso pueden ver las pinturas desde el bosque de robles sin entrar en el edificio.

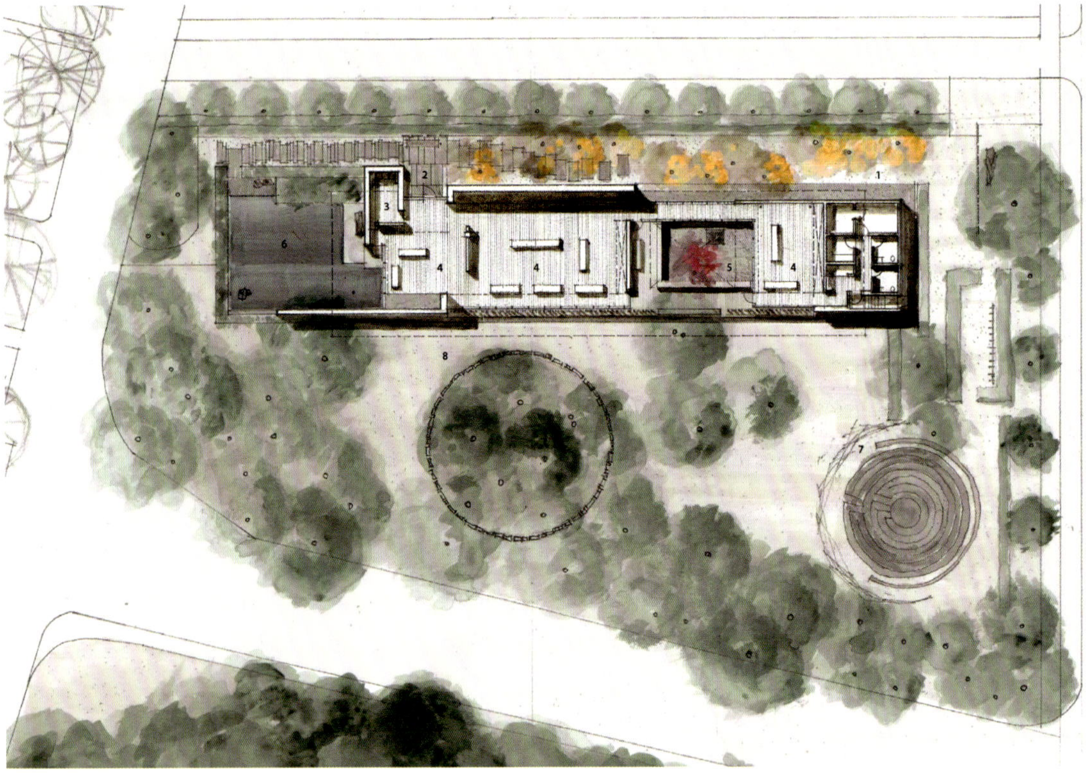

1. Entry garden
2. Entry
3. Information vestibule
4. Gallery
5. Exterior courtyard
6. Reflecting pool
7. Meditation labyrinth
8. Contemplative grove

Site plan

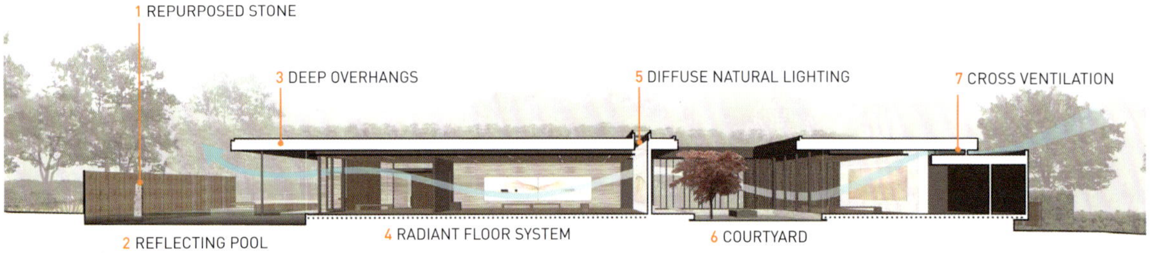

Sustainable strategies - south/north section

1. **Repurposed stone:** sculpture and exterior benches are fabricated from stone salvagedfrom the campus scrapyard.
2. **Reflecting pool:** shallow water element provides natural cooling for the southern exterior seating areas.
3. **Passive solar design:** Deep roof overhangs and insulated low-e coated glazing significantly reduce solar heat gain.
4. **Radiant heating and cooling:** radiant floor system provides heating and cooling for the building to give efficient comfort without noise or vibration.
5. **Filtered daylight:** skylights with motorized louvers and translucent fabric indirectly light the paintings, eliminating the need for artificial lighting during daylight hours and minimizing solar heat gain.
6. **Courtyard:** open courtyard provides additional light into the north gallery as well as to the bridge.
7. **Cross ventilation:** outside air is brought from the north and naturally travels to the south of the building where it is released.

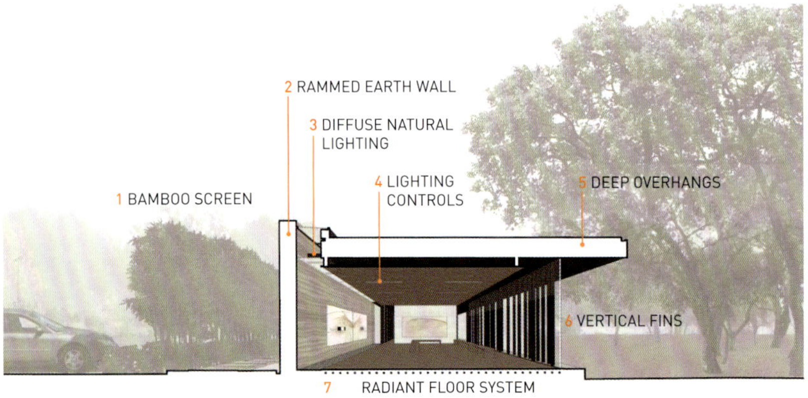

Sustainable strategies - west/east section

1. **Landscape screening:** continous hedge of tall bamboo provides additional screening to western heat gain.
2. **Earthen mass walls:** rammed earth walls provide a high thermal mass for temperature stabilization and passive cooling. Walls are made with soil excavated from the site, minimizing off-haul and reducing the use of off-site resources.
3. **Filtered daylight:** skylights with motorized louvers and traslucent fabric indirectly light the paintings, eliminating the need for artificial lighting during daylight hours and minimizing solar heat gain.
4. **Lighting controls:** timing controls and occupancy sensors adjust lighting levels to minimize energy usage.
5. **Passive solar design:** orientation of glazed walls to the oak grove to the east, deep roof overhangs, and insulated low-e glazing reduce solar heat gain.
6. **Vertical shading fins:** exterior metal fins along the length of glazing provide an integral shading element and minimize solar heat gain.
7. **Radiant heating and cooling:** radiant floor system provides heating and cooling for the building to give efficient comfort without noise or vibration.

ARKIN TILT ARCHITECTS

Founded in 1997, ATA is a Northern California firm whose award-winning projects embody a marriage of thoughtful design and ecology, creating spaces that are both comfortable and lyrical. Using primarily naturally low-carbon and carbon-storing materials —earth and straw in particular– they work to seamlessly integrate the natural and built environment. With a focus on craft and collaboration, ATA had the pleasure of working with the late David Easton of Rammed Earth Works on the Tip-Top Residence, and with rammed earth specialist Herman Bojarsky on the Hidden Villa Hostel and Summer Camp —inspiration for the masonry heater and kiva fireplace in the Canyon Road Bridge House— and more recently with Herman's son Jesse Bojarsky on a project with prefabricated rammed earth panels —another Easton innovation.

Gegründet im Jahr 1997 ist ATA eine Firma in Nordkalifornien, deren preisgekrönte Projekte eine Verbindung von durchdachtem Design und Ökologie darstellen und Räume schaffen, die sowohl komfortabel als auch lyrisch sind. Sie setzen hauptsächlich auf natürliche Materialien mit niedrigem Kohlenstoffgehalt und Kohlenstoffspeicherung - insbesondere Erde und Stroh - und arbeiten daran, die natürliche und gebaute Umgebung nahtlos zu integrieren. Mit Schwerpunkt auf Handwerk und Zusammenarbeit hatte ATA das Vergnügen, mit dem verstorbenen David Easton von Rammed Earth Works an der Tip-Top Residence zu arbeiten, mit Rammed Earth-Spezialist Herman Bojarsky an der Hidden Villa Hostel und Summer Camp - Inspiration für den Kaminofen und Kiva-Kamin im Canyon Road Bridge House - und neuerdings mit Hermans Sohn Jesse Bojarsky an einem Projekt mit vorgefertigten Rammed-Earth-Platten - einer weiteren Easton-Innovation.

CANYON ROAD BRIDGE HOUSE

Fondée en 1997, ATA est une entreprise de Californie du Nord dont les projets primés incarnent un mariage de design réfléchi et d'écologie, créant des espaces à la fois confortables et lyriques. En utilisant principalement des matériaux naturellement à faible teneur en carbone et stockant du carbone, tels que la terre et la paille en particulier, ils travaillent à intégrer de manière transparente l'environnement naturel et construit. Axée sur l'artisanat et la collaboration, ATA a eu le plaisir de travailler avec feu David Easton de Rammed Earth Works sur la résidence Tip-Top, avec le spécialiste de la terre battue Herman Bojarsky sur l'auberge Hidden Villa et le camp d'été - source d'inspiration pour le poêle de masse et la cheminée Kiva dans la maison de la Canyon Road Bridge – et plus récemment avec le fils de Herman, Jesse Bojarsky, sur un projet avec des panneaux de terre battue préfabriqués – une autre innovation d'Easton.

Fundada en 1997, ATA es una empresa de California del Norte cuyos proyectos galardonados encarnan un matrimonio de diseño reflexivo y ecología, creando espacios que son tanto cómodos como líricos. Utilizando principalmente materiales naturalmente bajos en carbono y almacenadores de carbono, como la tierra y la paja en particular, trabajan para integrar sin problemas el entorno natural y construido. Con un enfoque en la artesanía y la colaboración, ATA tuvo el placer de trabajar con el difunto David Easton de Rammed Earth Works en la Residencia Tip-Top, y con el especialista en tierra apisonada Herman Bojarsky en el Hostal Hidden Villa y el Campamento de Verano —inspiración para la estufa de mampostería y la chimenea kiva en la Casa del Puente de la Carretera del Cañón— y más recientemente con el hijo de Herman, Jesse Bojarsky, en un proyecto con paneles prefabricados de tierra apisonada —otra innovación de Easton.

WATERSHED STRAWBALE RESIDENCE

CANYON ROAD BRIDGE HOUSE

SANTA FE, NEW MEXICO, UNITED STATES
PHOTOS © KATE RUSSELL PHOTOGRAPHY

AWARDS:

- LEED Platinum

Design Architect: David Arkin, AIA
Design Project Manager: Devin Kinney
Client: Julie Stachowiak and Tuck Stibich
Contractor: Prull Custom Builders of Santa Fe
Drawings by: Arkin Tilt Architects
Landscape Architect: Donna Bone/Design with Nature
Civil Engineer: Morey Walker

This house in Santa Fe's historic district is located on a rolling wooded site, the dining room bridges a dry arroyo to access the living space which enjoys views of downtown Santa Fe. The house combines traditional construction elements with modern high-performance systems and richly textured materials. The living room features ICF walls with lime plaster, concrete floors, and a rammed earth masonry fireplace wall that adds thermal mass and improves the home's passive solar performance. Salvaged wood and locally harvested spruce are used for interior accents. Access to the house is through a covered courtyard with an outdoor rammed earth "kiva" - a take on the traditional Santa Fe fireplace. Sean Kaltenbach of New Mexico Earthworks led the construction of the rammed earth features.

Dieses Haus im historischen Viertel von Santa Fe liegt auf einem bewaldeten, hügeligen Grundstück. Das Esszimmer überquert einen trockenen Arroyo, um den Wohnraum zu erreichen, von dem aus man einen Blick auf die Innenstadt von Santa Fe hat. Das Haus kombiniert traditionelle Bauelemente mit modernen, leistungsstarken Systemen und reich strukturierten Materialien. Das Wohnzimmer verfügt über ICF-Wände mit Kalkputz, Betonböden und eine Wand aus Stampflehm-Mauerwerk, die thermische Masse hinzufügt und die passive Solarleistung des Hauses verbessert. Für die Innenausstattung werden wiedergewonnenes Holz und örtlich geerntete Fichte verwendet. Der Zugang zum Haus erfolgt über einen überdachten Innenhof mit einem Außenkamin aus Stampflehm - eine moderne Interpretation des traditionellen Santa Fe-Kamins. Sean Kaltenbach von New Mexico Earthworks leitete den Bau der Stampflehm-Merkmale.

Cette maison située dans le quartier historique de Santa Fe est située sur un terrain boisé vallonné, la salle à manger enjambe un arroyo sec pour accéder à l'espace de vie qui offre une vue sur le centre-ville de Santa Fe. La maison combine des éléments de construction traditionnels avec des systèmes modernes haute performance et des matériaux texturés riches. Le salon dispose de murs en béton coulé avec un enduit à la chaux, des sols en béton et un mur de cheminée en maçonnerie de terre compressée qui ajoute de la masse thermique et améliore les performances solaires passives de la maison. Du bois récupéré et de l'épicéa récolté localement sont utilisés pour les accents intérieurs. L'accès à la maison se fait par une cour couverte avec un « kiva » en terre compressée en extérieur - une version du foyer traditionnel de Santa Fe. Sean Kaltenbach de New Mexico Earthworks a dirigé la construction des éléments de terre compressée.

Esta casa en el distrito histórico de Santa Fe está ubicada en un terreno boscoso ondulado, el comedor conecta un arroyo seco para acceder al espacio habitable que goza de vistas del centro de Santa Fe. La casa combina elementos de construcción tradicionales con sistemas modernos de alto rendimiento y materiales texturizados de alta calidad. La sala de estar presenta paredes de ICF con yeso de cal, pisos de concreto y una pared de chimenea de albañilería de tierra apisonada que agrega masa térmica y mejora el rendimiento solar pasivo de la casa. Se utilizan madera recuperada y abeto cosechado localmente para los acentos interiores. El acceso a la casa es a través de un patio cubierto con una «kiva» de tierra apisonada al aire libre, una versión de la tradicional chimenea de Santa Fe. Sean Kaltenbach de New Mexico Earthworks lideró la construcción de las características de tierra apisonada.

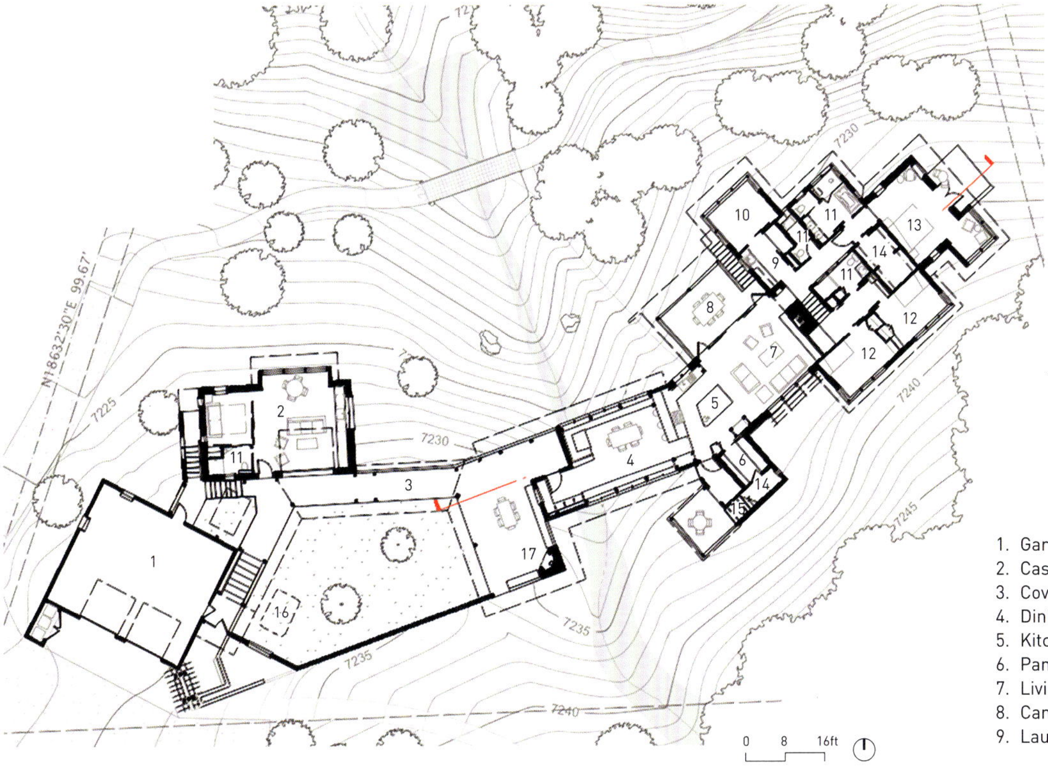

1. Garage
2. Casita-guest house
3. Covered walkway
4. Dining & entry bridge
5. Kitchen
6. Pantry
7. Living
8. Cantilevered deck
9. Laundry
10. Art studio
11. Bathroom
12. Bedroom
13. Master bedroom
14. Closet/storage
15. Stairs to roof
16. Stormwater cistern
17. KIVA

North elevation

WATERSHED STRAWBALE RESIDENCE

SONOMA COUNTY, CALIFORNIA, UNITED STATES
PHOTOS © EDWARD CALDWELL PHOTOGRAPHY

AWARDS:

- BUILD Architecture Awards 2016
- Recognized Leaders in Low Carbon Building, USA & Best Californian Off-Grid Home Design
- Custom Builder "High Performance Homes," July 2016
- Vetrazzo Design Star, 2016

Design Architect: David Arkin, AIA
Design Project Manager: Amanda Knowles
Client: Withheld, by owner's request
Drawings by: Arkin Tilt Architects
Structural Engineer: Verdant Structural Engineers
Landscape Architect: Angelica Papio
Civil Engineer: Atterbury & Associates

This off-grid homestead was carefully designed and constructed to minimize impact on the surrounding landscape. The home's indoor/outdoor flow creates a tranquil nature-centered living space for the family. One of the first projects to use David Easton's compressed earth block (now known as EcoTerra block), the masonry walls deliveries at the remote site provide a warm, textured feel. The upper walls are made of highly-insulating straw bale, covered with lime plaster for thermal mass. Much of the timber used in the home was milled on site, including live-edge slabs used as a guardrail at the stairway to the bedrooms. The living room roof, planted with vegetation, blends seamlessly with the landscape. Other roofs, made of gently curving corrugated metal, are soften version of typical agrarian structure while mimicing the surrounding tree canopies. Overall, this residence showcases innovative passive solar design with a hybrid palette of sustainable materials.

Diese autarke Hofstelle wurde sorgfältig entworfen und gebaut, um die Auswirkungen auf die umgebende Landschaft zu minimieren. Der Indoor-Outdoor-Fluss des Hauses schafft einen ruhigen Naturraum für die Familie. Eines der ersten Projekte, die David Eastons verdichteten Erdblock (jetzt bekannt als EcoTerra Block) verwendeten, lieferten die Mauerwerksmauern am entlegenen Standort ein warmes, texturiertes Gefühl. Die oberen Wände sind aus hochisolierendem Strohballen gefertigt und mit Kalkputz für die thermische Masse bedeckt. Ein Großteil des Holzes im Haus wurde vor Ort gesägt, einschließlich Live-Edge-Platten, die als Geländer am Treppenaufgang zu den Schlafzimmern verwendet wurden. Das Wohnzimmerdach, das mit Vegetation bepflanzt ist, fügt sich nahtlos in die Landschaft ein. Andere Dächer, die aus sanft gebogenem Wellblech bestehen, sind eine weichere Version typischer Agrarstrukturen und imitieren die umliegenden Baumkronen. Insgesamt präsentiert diese Residenz innovative und nachhaltige Materialien.

Cette ferme autonome a été soigneusement conçue et construite pour minimiser l'impact sur le paysage environnant. Le flux intérieur / extérieur de la maison crée un espace de vie tranquille centré sur la nature pour la famille. L'un des premiers projets à utiliser le bloc de terre comprimée de David Easton (maintenant connu sous le nom de bloc EcoTerra), les murs de maçonnerie livrés sur le site distant fournissent une sensation chaleureuse et texturée. Les murs supérieurs sont faits de paille hautement isolante, recouverts de plâtre de chaux pour la masse thermique. Une grande partie du bois utilisé dans la maison a été sciée sur place, y compris des plaques de bord vivant utilisées comme garde-corps à l'escalier menant aux chambres. Le toit du salon, planté de végétation, se fond parfaitement dans le paysage. D'autres toits, faits de tôles ondulées légèrement incurvées, sont une version plus douce de la structure agraire typique tout en imitant les canopées des arbres environnants. Dans l'ensemble, cette résidence présente une palette hybride de matériaux durables et innovants.

Este hogar fuera de la red fue cuidadosamente diseñado y construido para minimizar el impacto en el paisaje circundante. El flujo interior/exterior de la casa crea un espacio de vida tranquilo y centrado en la naturaleza para la familia. Uno de los primeros proyectos en utilizar el bloque de tierra comprimido de David Easton (ahora conocido como bloque EcoTerra), las paredes de mampostería entregadas en el sitio remoto proporcionan una sensación cálida y texturizada. Las paredes superiores están hechas de paja altamente aislante, cubiertas con yeso de cal para la masa térmica. Gran parte de la madera utilizada en la casa se cortó en el sitio, incluidas las losas de borde vivo utilizadas como barandilla en la escalera hacia los dormitorios. El techo de la sala de estar, plantado con vegetación, se fusiona perfectamente con el paisaje. Otros techos, hechos de láminas corrugadas de metal suavemente curvadas, son una versión más suave de la estructura agraria típica mientras imitan las copas de los árboles circundantes. En general, esta residencia muestra una paleta híbrida de materiales sostenibles e innovadores.

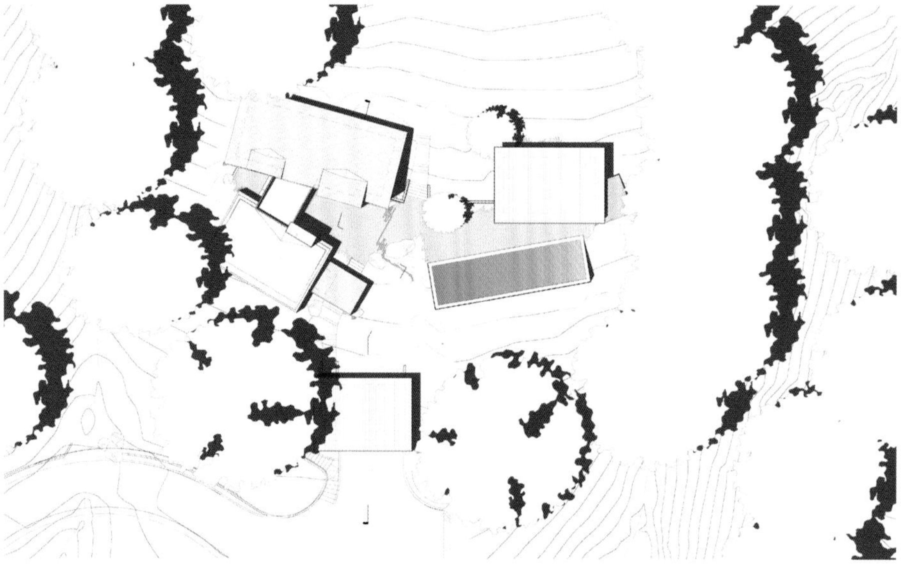

Site and roof plan

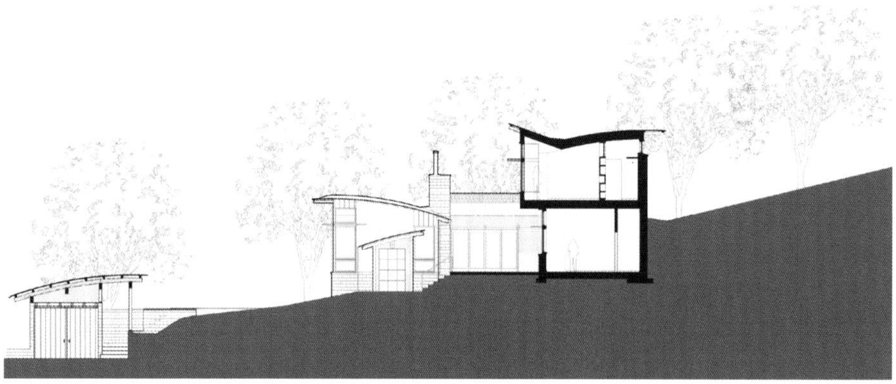

Section

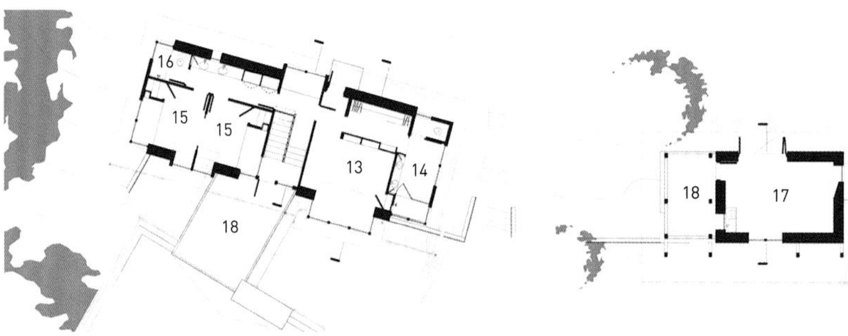

Upper level plan

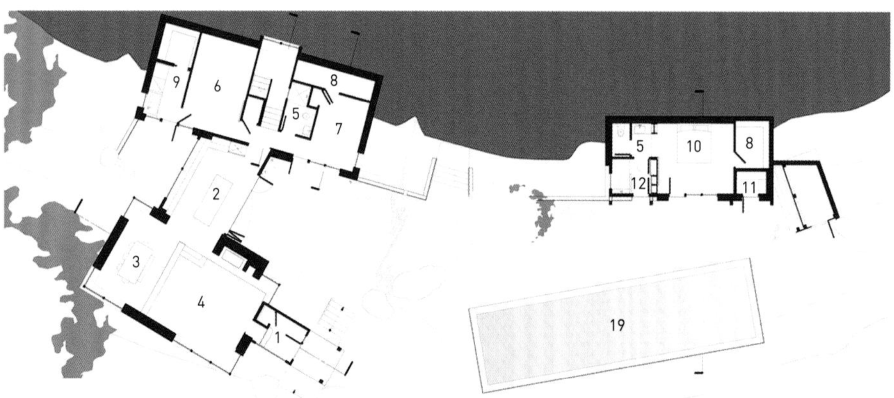

Lower level plan

0 8 16ft

1. Entry/mudroom
2. Kitchen
3. Dining
4. Living
5. Bathroom
6. Media room
7. Office
8. Storage/ mechanical
9. Laundry
10. Pool house
11. Sauna
12. Changing room
13. Master bedroom
14. Master bathroom
15. Kid's bedroom
16. Kid's bathroom
17. Art room
18. Roof deck
19. Pool

ASQUARED STUDIOS // JESSIE WHITESIDES, NCARB - TONY GARCIA, AIA

Asquared is a contemporary design studio seeking to embrace projects that present us with the challenges of vernacular living within the context of our modern environment. We believe that our clients come first, and that all projects require an interactive collaboration to develop an artfully crafted result. We are invigorated by the notion that the basic fundamentals of function and cost can be translated into sophisticated spaces, elegant details, and environmental sustainability.

By offering and providing a principal-led delivery system we have taken great care to shape our practice into a uniquely responsive and personable studio design experience. Throughout our careers, we have successfully collaborated with a wide range of clients to satisfy diverse sets of goals. Our projects include residential, winery and hospitality, commercial, and interior improvement projects.

Our wide-ranging portfolio is rooted in a never ending curiosity in how we live and work today, and the pleasure we take in collectively solving design challenges through research, strategic thinking, spatial analysis and a focus on craft.

With studios located in San Diego and Sonoma County, California, we are able to identify with our constantly developing modern world.

Asquared ist ein zeitgenössisches Designstudio, das sich auf Projekte einlässt, die uns vor die Herausforderungen des traditionellen Lebens im Kontext unserer modernen Umwelt stellen. Wir glauben, dass unsere Kunden an erster Stelle stehen und dass alle Projekte eine interaktive Zusammenarbeit erfordern, um ein kunstvoll gestaltetes Ergebnis zu entwickeln. Der Gedanke, dass die grundlegenden Prinzipien von Funktion und Kosten in anspruchsvolle Räume, elegante Details und ökologische Nachhaltigkeit umgesetzt werden können, beflügelt uns.

Durch das Angebot und die Bereitstellung eines von den Auftraggebern geleiteten Systems haben wir darauf geachtet, unsere Praxis zu einer einzigartig reaktionsschnellen und sympathischen Studio-Design-Erfahrung zu machen. Im Laufe unserer Karriere haben wir erfolgreich mit einer Vielzahl von Kunden zusammengearbeitet, um die unterschiedlichsten Ziele zu erreichen. Unsere Projekte umfassen Wohnhäuser, Weingüter und Gaststätten, gewerbliche Projekte und Projekte zur Innenraumgestaltung.

Unser breit gefächertes Portfolio wurzelt in der unermüdlichen Neugier auf die Art und Weise, wie wir heute leben und arbeiten, und in der Freude, die wir bei der gemeinsamen Lösung von Designherausforderungen durch Forschung, strategisches Denken, räumliche Analyse und den Fokus auf das Handwerk empfinden.

Mit Studios in San Diego und Sonoma County, Kalifornien, sind wir in der Lage, uns mit unserer sich ständig entwickelnden modernen Welt zu identifizieren.

SONOMA RESIDENCE

Asquared est un studio de design contemporain qui cherche à s'engager dans des projets qui nous présentent les défis de la vie vernaculaire dans le contexte de notre environnement moderne. Nous pensons que nos clients passent avant tout, et que tous les projets nécessitent une collaboration interactive pour aboutir à un résultat artistiquement élaboré. Nous sommes stimulés par l'idée que les principes fondamentaux de la fonction et du coût peuvent être traduits en espaces sophistiqués, en détails élégants et en durabilité environnementale.
En proposant et en fournissant un système de livraison dirigé par le directeur, nous avons pris grand soin de façonner notre pratique en une expérience de conception de studio unique, réactive et personnelle. Tout au long de notre carrière, nous avons collaboré avec succès avec un large éventail de clients pour satisfaire des objectifs variés. Nos projets comprennent des projets résidentiels, vinicoles et hôteliers, commerciaux et d'amélioration intérieure.
Notre vaste portefeuille est ancré dans une curiosité sans fin pour la façon dont nous vivons et travaillons aujourd'hui, et le plaisir que nous prenons à résoudre collectivement les défis de la conception par la recherche, la réflexion stratégique, l'analyse spatiale et l'accent mis sur l'artisanat.
Avec des studios situés à San Diego et dans le comté de Sonoma, Californie, nous sommes en mesure de nous identifier à notre monde moderne en constante évolution.

Asquared es un estudio de diseño contemporáneo que busca abarcar proyectos que nos presenten los retos de la vida vernácula dentro del contexto de nuestro entorno moderno. Creemos que nuestros clientes son lo primero, y que todos los proyectos requieren una colaboración interactiva para desarrollar un resultado artísticamente elaborado. Nos anima la idea de que los fundamentos básicos de funcionalidad y coste pueden traducirse en espacios sofisticados, detalles elegantes y sostenibilidad medioambiental.
Al ofrecer y proporcionar un sistema de entrega dirigido por el director, hemos tenido mucho cuidado para dar forma a nuestra práctica en una experiencia de diseño de estudio singularmente sensible y personal. A lo largo de nuestras carreras, hemos colaborado con éxito con una amplia gama de clientes para satisfacer diversos objetivos. Nuestros proyectos incluyen proyectos residenciales, de bodegas y hostelería, comerciales y de mejora de interiores.
Nuestra amplia cartera se basa en una curiosidad sin fin en la forma en que vivimos y trabajamos hoy en día, y el placer que tomamos en la solución colectiva de los desafíos de diseño a través de la investigación, el pensamiento estratégico, el análisis espacial y un enfoque en la artesanía.
Con estudios ubicados en San Diego y el condado de Sonoma, California, somos capaces de identificarnos con nuestro mundo moderno en constante desarrollo.

SONOMA RESIDENCE

SONOMA COUNTY, CALIFORNIA, UNITED STATES
PHOTOS © ADAM POTTS PHOTOGRAPHY, ADAMPOTTSPHOTOGRAPHY.COM,
IG: @ADAMPOTTSPHOTO

AWARDS:

2022 AIA Redwood Empire Merit Award

Design Architect: Asquared Studios, Jessie Whitesides, Ncarb, Architect Tony Garcia, AIA Architect
Client: Robert Mark Kamen
Landscape Architect: Roche + Roche
General Contractor: Annadel Builders, Inc.

Influenced by Japanese pagoda roof forms, this single-story vineyard residence uses a textural material palette of golden rammed earth, corten steel, renewable wood, and high-performing glass. Designed to integrate itself into the rugged terrain, the design was developed as a series of interlocking boxes, arranged to maximize the views. The rammed earth walls frame windswept trees and seem to map the rows of grapevines renowned for making Kamen Estate Wines. Corten steel siding defines the structure's distinct forms. A pagoda-style roof elevates the ceiling line above the expanses of window walls, opening the home to the land. Inside, the rammed earth is accentuated against neutral paint colors and accents of walnut, *shou sugi ban*, blackened steel, and concrete floors finished in a unique distressed finish.
This home is significant to the history of rammed earth construction in the United States, as it was the last true rammed earth home built by Rammed Earthworks and David Easton, the Grandfather of modern rammed earth construction.

Beeinflusst von japanischen Pagodendachformen, verwendet dieses einstöckige Wohnhaus in den Weinbergen eine strukturierte Materialpalette aus goldenem Stampflehm, Cortenstahl, erneuerbarem Holz und hochleistungsfähigem Glas. Um sich in das zerklüftete Gelände zu integrieren, wurde das Design als eine Reihe von ineinandergreifenden Boxen entwickelt, die so angeordnet sind, dass die Aussicht maximiert wird. Die Stampflehmwände rahmen windgepeitschte Bäume ein und scheinen die Reihen von Weinstöcken abzubilden, die für die Herstellung der Kamen Estate Wines bekannt sind. Cortenstahlverkleidungen definieren die unterschiedlichen Formen der Struktur. Ein pagodenartiges Dach erhebt die Deckenlinie über die weiten Fensterwände und öffnet das Haus zum Land hin. Im Inneren wird die Stampflehmkonstruktion durch neutrale Farbtöne und Akzente aus Nussbaumholz, Shou Sugi Ban, geschwärztem Stahl und Betonböden mit einzigartigem Distressed-Finish akzentuiert.
Dieses Haus ist für die Geschichte des Stampflehmbaus in den Vereinigten Staaten von Bedeutung, da es das letzte echte Stampflehmhaus war, das von Rammed Earthworks und David Easton, dem Großvater des modernen Stampflehmbaus, gebaut wurde.

Influencée par les formes japonaises des toits en pagode, cette résidence viticole d'un étage utilise une palette de matériaux texturés composée de terre battue dorée, d'acier corten, de bois renouvelable et de verre à haute performance. Conçue pour s'intégrer dans le terrain accidenté, la conception a été développée comme une série de boîtes imbriquées, disposées de manière à maximiser les vues. Les murs en pisé encadrent les arbres balayés par le vent et semblent dessiner les rangées de vignes réputées pour la production des vins du domaine Kamen. Le bardage en acier Corten définit les formes distinctes de la structure. Un toit en forme de pagode élève la ligne de plafond au-dessus des étendues de fenêtres, ouvrant la maison sur la terre. À l'intérieur, la terre battue est mise en valeur par des couleurs de peinture neutres et des accents de noyer, de shou sugi ban, d'acier noirci et de sols en béton avec une finition unique.
Cette maison est importante pour l'histoire de la construction en pisé aux États-Unis, car il s'agit de la dernière véritable maison en pisé construite par Rammed Earthworks et David Easton, le grand-père de la construction moderne en pisé.

Influenciada por las formas japonesas de los tejados de una pagoda, esta residencia de una sola planta en un viñedo utiliza una paleta de materiales texturizados de tierra apisonada dorada, acero corten, madera renovable y vidrio de alto rendimiento. Diseñado para integrarse en el accidentado terreno, el diseño se desarrolló como una serie de cajas entrelazadas, dispuestas para maximizar las vistas. Los muros de tierra apisonada enmarcan árboles azotados por el viento y parecen trazar las hileras de vides famosas por la elaboración de los vinos Kamen Estate. El revestimiento de acero corten define las distintas formas de la estructura. Un tejado en forma de pagoda eleva la línea del techo por encima de las grandes ventanas, abriendo la casa a la tierra. En el interior, la tierra apisonada se acentúa con colores neutros de pintura y detalles de nogal, shou sugi ban, acero ennegrecido y suelos de hormigón con un singular acabado envejecido.
Esta casa es importante para la historia de la construcción con tierra apisonada en los Estados Unidos, ya que fue la última verdadera casa de tierra apisonada construida por Rammed Earthworks y David Easton, el abuelo de la construcción moderna con tierra apisonada.

Site plan

- Design axis: arrival progression + revealed view to San Francisco Bay
- A. Heritage Oaks provide shade + support under-story growth attracting native pollinators
- B. Exterior walls become interior finish with excellent thermal mass. Rammed Earth is a Low-carbon + Fire-resistant material
- C. Native and drought tolerant landscaping surrounds the home bringing the outdoors - in
- D. The house is composed of interlocking boxes designed to maximize daylighting and views
- E. Sliding window wall opens to outdoor seating area, extending the interior living space outdoors
- F. Estate vineyards

1. Entry
2. Dining
3. Living
4. Kitchen
5. Master wing
6. Zen garden
7. Closet
8. Wine library
9. Laundry
10. Gallery hall
11. Yoga room
12. Powder room
13. Office
14. Outdoor living

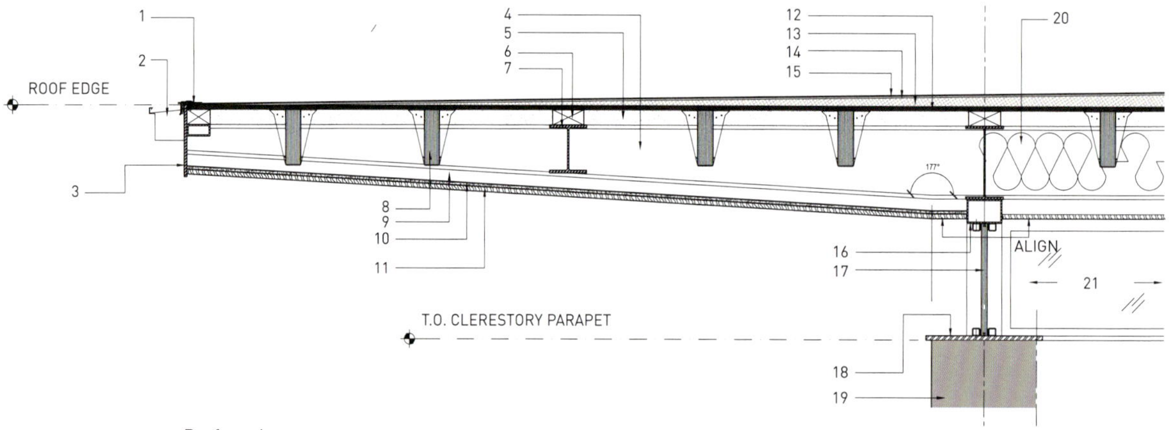

Roof overhang

1. Durolast prefabricated drip edge with factory welded skirt and rectangular gutter.
2. Box gutter in corten finish to match nosing plate
3. 1/2" x 12" steel nosing plate, corten finish
4. Structural steel girder beyond, S.S.D.
5. 2 1/2" Closed cell spray foam insulation to fill nailer space
6. 3x nailer typical, S.S.D.
7. Structural steel framing, S.S.D.
8. LVL framing, S.S.D., typical
9. Provide furring to dropped wood ceiling
10. 5/8" type 'x' dry wall as backing to wood ceiling finish
11. 1x8 wood plank ceiling, made from 1 x 10 board with 1.25" Shiplap joint. See finish schedule for additional information. Joint to be set tight see rcp for board layout
12. 5/8" Plywood shear, S.S.D.
13. Sloped rigid insulation, 1/8" per ft.
14. 1/4" Densdeck roofing board
15. Class 'a' roofing: 50 mil. Durolast membrane
16. HSS 6" x 4" as top chord of vierendeel truss, S.S.D.
17. Clerestory window
18. 4" Steel plate as bottom chord of vierendeel truss, S.S.D.
19. Rammed earth wall
20. Provide R30 batt insulation at interior
21. Clerestory window at side wall

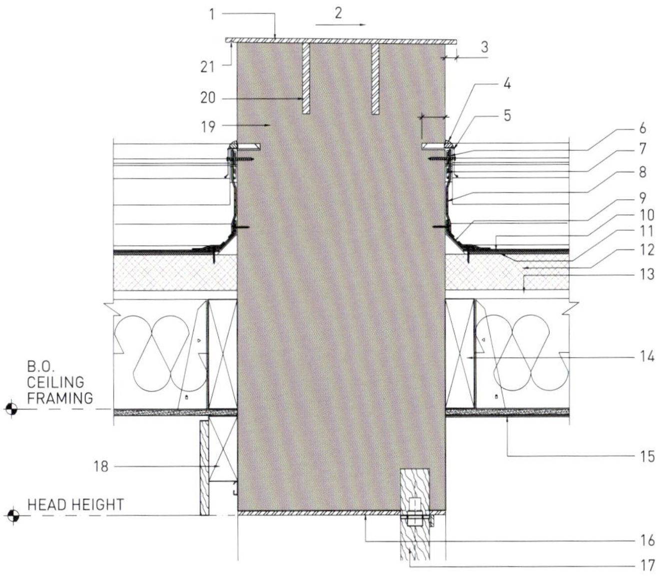

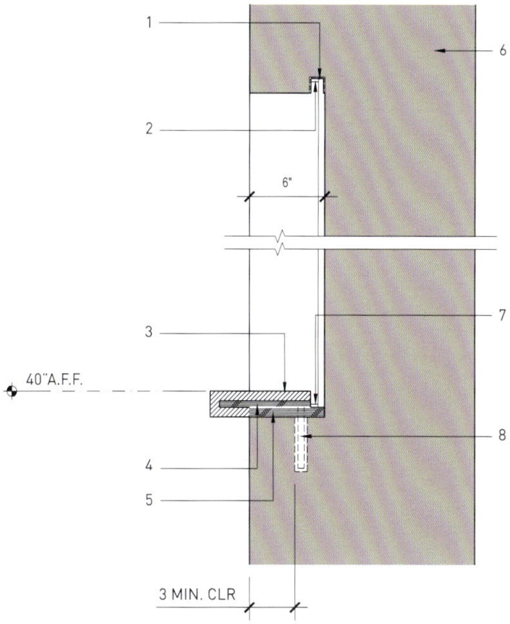

Parapet detail

1. 3/8" Corten steel plate
2. Slope down 1/8" per ft
3. 1" Each side
4. Duro-last chaulk
5. Stainless steel counter-flashing fastened 6-inches on center with duro-last fastener. Let-in to rammed earth wall
6. Duro-last membrane must extend 3/4-inch beyond fastener
7. Duro-last sealant
8. Duro-last prefabricated parapet wall flashing
9. Durolast cant strip
10. Durolast 50mil roofing
11. 1/4" Dens deck roof board
12. Sloped foam panel
13. Roof sheathing, S.S.D.
14. Roof framing and insulation per plans, S.S.D.
15. Gyp. Board finish per plans
16. See door detail at this location for additional information
17. Door per schedule
18. 3/4" x 8" wood Trim and picture rail, typ.
19. Rammed earth wall per plan
20. 5/8" x 6" steel dowels welded to plate at 24" O.C. (Staggered)
21. 1/8" Radius drip edge

Nitch Detail

1. 1 1/4" x 1 1/4" x 1/8" aluminum u-channel embedded into rammed earth at head and jambs of niche, paint channel black
2. Led strip light cont. around niche set into aluminum channel
3. Solid surface niche shelf, miter corners typical
4. 1/2" P.T. Plywood
5. 3/4" P.T. Plywood
6. Rammed earth wall
7. Mount lighting on plywood shelf support at sill. Paint plywood black
8. Drill and epoxy attachment for shelf to rammed earth

AST77 ARCHITECTEN EN INGENIEURS // PETER VAN IMPE

AST 77 is a strongly locally committed architectural and engineering firm where architects, engineers and interior architects work together on high-quality projects.
The team has built up great expertise in design, research and technical analysis over the past 15 years. Contemporary and innovative solutions were developed for present-day assignments - varying in scale and nature - for both new construction and renovation and with different functions. The architecture developed is characterized by three key words: Context, Structure and Flexibility. Specific project-related aspects such as function, location, typology, social context, materialization, energy, environment and circularity are linked to these keywords. With this forward-looking view, AST 77 aroused the interest of national and international awards such as the Belgian Building Awards and Terra Fibra Award.

AST 77 ist ein stark lokal engagiertes Architektur- und Ingenieurbüro, in dem Architekten, Ingenieure und Innenarchitekten gemeinsam an hochwertigen Projekten arbeiten.
Das Team hat in den letzten 15 Jahren ein großes Fachwissen in den Bereichen Design, Forschung und technische Analyse aufgebaut. Es wurden zeitgemäße und innovative Lösungen für aktuelle Aufgabenstellungen - unterschiedlicher Größe und Art - sowohl für Neubau als auch für Renovierung und mit verschiedenen Funktionen entwickelt. Die entwickelte Architektur zeichnet sich durch drei Schlüsselbegriffe aus: Kontext, Struktur und Flexibilität. Mit diesen Schlüsselwörtern sind spezifische projektbezogene Aspekte wie Funktion, Standort, Typologie, sozialer Kontext, Materialisierung, Energie, Umwelt und Kreislaufwirtschaft verbunden. Mit dieser zukunftsweisenden Sichtweise hat AST 77 das Interesse nationaler und internationaler Auszeichnungen wie der Belgian Building Awards und Terra Fibra Award geweckt.

AST 77 est un cabinet d'architecture et d'ingénierie fortement engagé localement, où architectes, ingénieurs et architectes d'intérieur travaillent ensemble sur des projets de haute qualité.
L'équipe a acquis une grande expertise en matière de conception, de recherche et d'analyse technique au cours des 15 dernières années. Des solutions contemporaines et innovantes ont été développées pour des missions actuelles - d'échelle et de nature différentes - tant pour des constructions neuves que pour des rénovations et avec des fonctions différentes. L'architecture développée se caractérise par trois mots clés : Contexte, Structure et Flexibilité. Des aspects spécifiques liés au projet tels que la fonction, l'emplacement, la typologie, le contexte social, la matérialisation, l'énergie, l'environnement et la circularité sont liés à ces mots clés. Avec cette vision prospective, AST 77 a suscité l'intérêt de prix nationaux et internationaux tels que les Belgian Building Awards et Terra Fibra Award.

AST 77 es una empresa de arquitectura e ingeniería con un fuerte compromiso local en la que arquitectos, ingenieros y arquitectos de interiores trabajan juntos en proyectos de alta calidad.
El equipo ha acumulado una gran experiencia en diseño, investigación y análisis técnico durante los últimos 15 años. Se han desarrollado soluciones contemporáneas e innovadoras para los encargos actuales —de diferente escala y naturaleza— tanto para nuevas construcciones como para renovaciones y con diferentes funciones. La arquitectura desarrollada se caracteriza por tres palabras clave: Contexto, Estructura y Flexibilidad. A estas palabras clave se vinculan aspectos específicos del proyecto como la función, la ubicación, la tipología, el contexto social, la materialización, la energía, el medio ambiente y la circularidad. Con esta visión de futuro, AST 77 despertó el interés de premios nacionales e internacionales como los Belgian Building Awards y Terra Fibra Award.

A 15 METER HIGH RAMMED EARTH WALL

TIENEN, BELGIUM // PHOTOS © THOMASNOCETO, STEVEN MASSART, MAARTEN DE BOUW, PHILIPPE VAN DEN PANHUYZEN

Project Team: Peter Van Impe, Jordi Gijzen, Marijke Peeters, Evelien Broeckx
Total Built Area: 250 m²
General Contractor: Inter ID, Kestens Montage, Elektriciteitswerken Degens, Sollec, Ronan Marteau, Dakwerken Makkedak, Gert Smets, Air Service Center, Alu2+, Het Leemniscaat
Engineers: BC Studies, Util, Yuvico, 2B-Safe

Architect Peter Van Impe (AST77) wants to builds his own home in Tienen, Belgium, and is intrigued by contemporary Rammed Earth Construction. Together with BC studies, he proposes to use earth from the site, made available through the excavation of the semi-submerged basement. A design is conceived in which all spaces over 5 floors are organized around a single 15 meter high unstabilized rammed earth wall. Specimens where made to define colour, but more importantly, to achieve sufficient compression strengths of 3,2 MPa. For this, we mixed earth from the site with sand, gravels and pottery clay from Les Argilières Hins. We worked together onsite with contractor Het Leemniscaat for a smooth execution of works. A heating cable emitting 40°C heat was rammed into the wall for preventing frost, since the works took place in winter. In the end, a beautiful wall made of only soils stands tall in this innovative project, which could only happen through an open and experimental mindset of all partners: AST77, BC studies, Argilières Hins and Het Leemniscaat.

Architekt Peter Van Impe (AST77) möchte sein eigenes Haus in Tienen, Belgien, bauen und ist von zeitgenössischer Stampflehm-Konstruktion fasziniert. Zusammen mit BC Studies schlägt er vor, Erde von der Baustelle zu verwenden, die durch die Ausgrabung des halbversunkenen Kellers verfügbar wird. Ein Design wird entwickelt, bei dem alle Räume über 5 Stockwerke um eine einzelne 15 Meter hohe unstabilisierte Stampflehmwand organisiert sind. Proben wurden hergestellt, um die Farbe zu definieren, aber vor allem, um ausreichende Druckfestigkeiten von 3,2 MPa zu erreichen. Dafür mischten wir Erde von der Baustelle mit Sand, Kies und Töpfererde von Les Argilières Hins. Wir arbeiteten vor Ort zusammen mit dem Auftragnehmer Het Leemniscaat für eine reibungslose Ausführung der Arbeiten. Ein Heizkabel, das 40°C Wärme abgibt, wurde in die Wand gerammt, um Frost zu verhindern, da die Arbeiten im Winter stattfanden. Am Ende steht in diesem innovativen Projekt eine schöne Wand aus nur Boden, die nur durch eine offene und experimentelle Einstellung aller Partner: AST77, BC Studies, Argilières Hins und Het Leemniscaat entstehen konnte.

L'architecte Peter Van Impe (AST77) souhaite construire sa propre maison à Tienen, en Belgique, et est intrigué par la construction contemporaine en terre crue. Avec BC Studies, il propose d'utiliser la terre du site, mise à disposition grâce à l'excavation du sous-sol semi-immergé. Un design est conçu dans lequel tous les espaces sur 5 étages sont organisés autour d'un seul mur en terre crue non stabilisé de 15 mètres de haut. Des échantillons ont été réalisés pour définir la couleur, mais surtout pour atteindre des résistances à la compression suffisantes de 3,2 MPa. Pour cela, nous avons mélangé la terre du site avec du sable, des graviers et de l'argile à poterie de Les Argilières Hins. Nous avons travaillé sur place avec l'entrepreneur Het Leemniscaat pour une exécution fluide des travaux. Un câble chauffant émettant une chaleur de 40°C a été enfoncé dans le mur pour prévenir le gel, car les travaux ont eu lieu en hiver. Au final, un beau mur fait uniquement de terres se dresse dans ce projet innovant, qui n'aurait pu être réalisé que grâce à une mentalité ouverte et expérimentale de tous les partenaires: AST77, BC Studies, Argilières Hins et Het Leemniscaat.

El arquitecto Peter Van Impe (AST77) quiere construir su propia casa en Tienen, Bélgica, y está interesado en la construcción contemporánea de tierra apisonada. Junto con BC Studies, propone utilizar la tierra del sitio, disponible a través de la excavación del sótano semisumergido. Se concibe un diseño en el que todos los espacios de más de 5 plantas están organizados alrededor de una sola pared de tierra apisonada no estabilizada de 15 metros de altura. Se realizaron muestras para definir el color, pero lo más importante fue lograr resistencias a la compresión suficientes de 3,2 MPa. Para esto, mezclamos tierra del sitio con arena, gravilla y arcilla de cerámica de Les Argilières Hins. Trabajamos juntos en el sitio con el contratista Het Leemniscaat para una ejecución fluida de las obras. Se empotró un cable calefactor que emite calor de 40°C en la pared para prevenir la congelación, ya que los trabajos se realizaron en invierno. Al final, una hermosa pared hecha solo de tierra se eleva en este innovador proyecto, que solo podría haberse logrado a través de una mentalidad abierta y experimental de todos los socios: AST77, BC Studies, Argilières Hins y Het Leemniscaat.

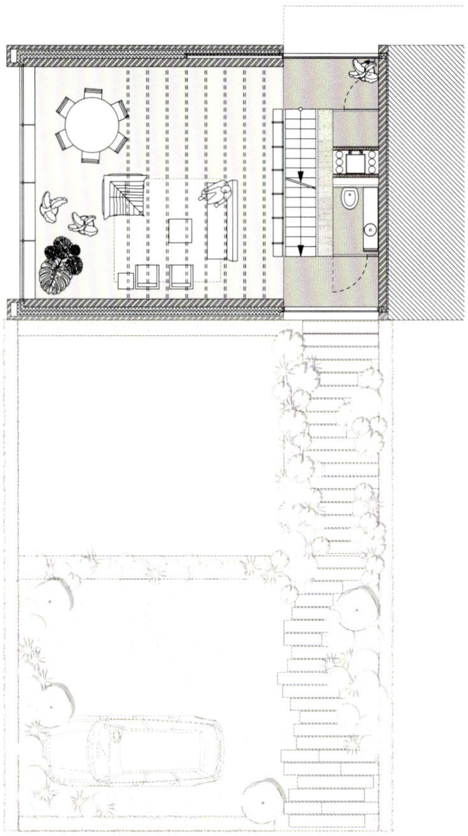

Second level plan

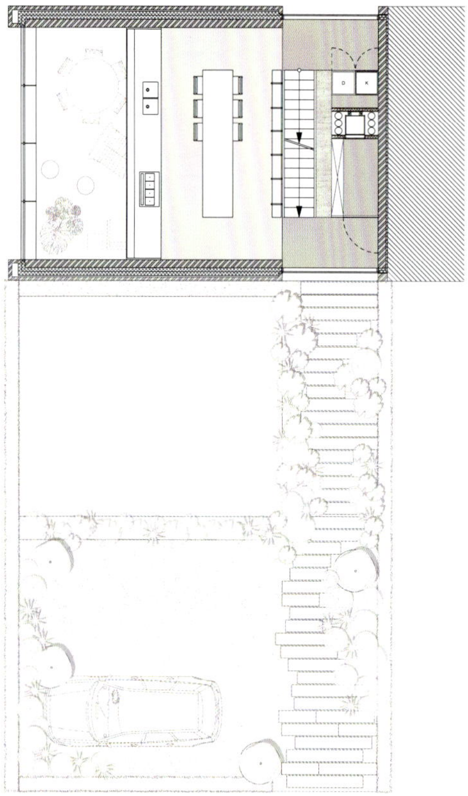

Third level plan

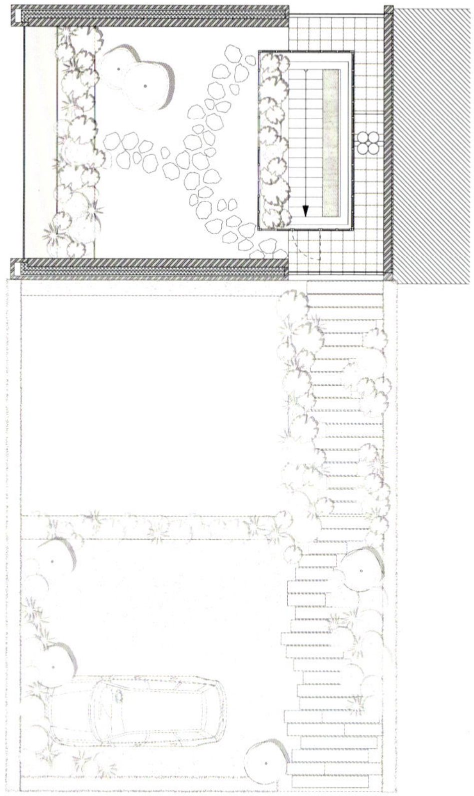

Roof plan

First level plan

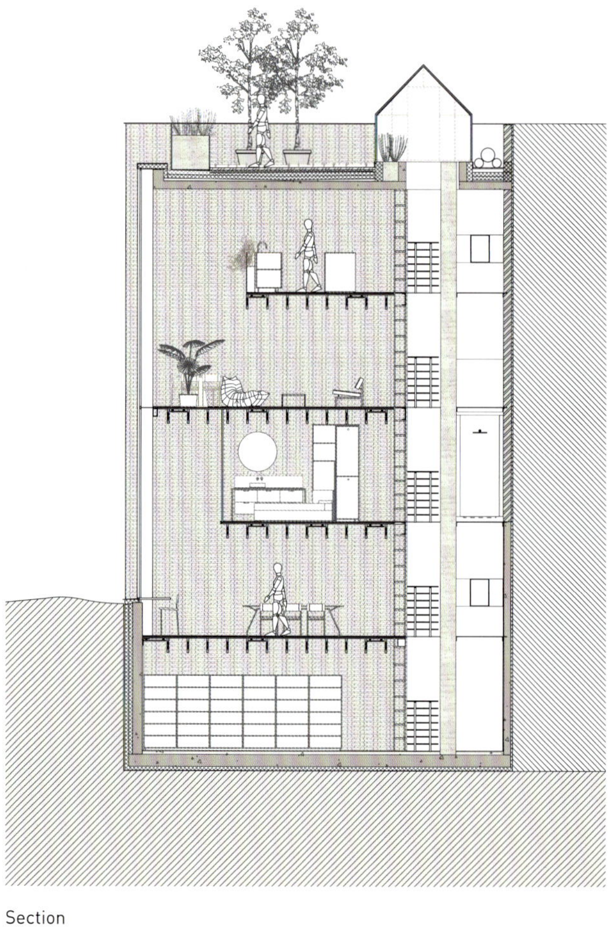

Section

GODTS

We are a multidisciplinary firm working on ecological architecture and intelligent water and sanitation designs since its inception in 1990. Through our work, we strive to showcase that an ecological design practice can be mainstream and relevant in various contexts, including geography, typology and scale, balanced by the client's aspirations for aesthetics, functionality and budget.
This effort has been especially focused on the urban scenario, using materials and systems to create a language of action that is not bound by preconceptions of what defines architecture. Every project is a test bed for developing this language, using a collaborative multidisciplinary approach grounded in informed decisions.

Wir sind ein multidisziplinäres Büro, das sich seit seiner Gründung im Jahr 1990 mit ökologischer Architektur und intelligentem Wasser- und Sanitärdesign beschäftigt. Mit unserer Arbeit wollen wir zeigen, dass ökologisches Design in verschiedenen Kontexten, einschließlich Geografie, Typologie und Maßstab, in einem ausgewogenen Verhältnis zu den Ansprüchen des Kunden an Ästhetik, Funktionalität und Budget stehen kann.
Diese Bemühungen konzentrieren sich vor allem auf das urbane Szenario, wobei Materialien und Systeme verwendet werden, um eine Handlungssprache zu schaffen, die nicht durch vorgefasste Meinungen darüber, was Architektur ausmacht, gebunden ist. Jedes Projekt ist ein Prüfstand für die Entwicklung dieser Sprache, wobei ein gemeinschaftlicher, multidisziplinärer Ansatz verwendet wird, der auf fundierten Entscheidungen beruht.

HAMSA'S HOUSE

Nous sommes un cabinet multidisciplinaire qui travaille sur l'architecture écologique et les conceptions intelligentes de l'eau et de l'assainissement depuis sa création en 1990. Par notre travail, nous nous efforçons de démontrer qu'une pratique de conception écologique peut être courante et pertinente dans divers contextes, y compris la géographie, la typologie et l'échelle, équilibrée par les aspirations du client en matière d'esthétique, de fonctionnalité et de budget.
Cet effort s'est particulièrement concentré sur le scénario urbain, en utilisant des matériaux et des systèmes pour créer un langage d'action qui n'est pas lié à des idées préconçues sur ce qui définit l'architecture. Chaque projet est un banc d'essai pour développer ce langage, en utilisant une approche multidisciplinaire collaborative fondée sur des décisions éclairées.

Somos una empresa multidisciplinar que trabaja en arquitectura ecológica y diseños inteligentes de agua y saneamiento desde su creación en 1990. A través de nuestro trabajo, nos esforzamos por demostrar que un estudio de diseño ecológico puede ser habitual y relevante en diversos contextos, como la geografía, la tipología y la escala, equilibrada por las aspiraciones del cliente en cuanto a estética, funcionalidad y presupuesto. Este esfuerzo se ha centrado especialmente en el escenario urbano, utilizando materiales y sistemas para crear un lenguaje de actuación que no esté sujeto a ideas preconcebidas sobre lo que define la arquitectura. Cada proyecto es un banco de pruebas para desarrollar este lenguaje, utilizando un enfoque multidisciplinar de colaboración basado en decisiones fundamentadas.

WAGHOBA ECOLODGE

HAMSA'S HOUSE

R. T. NAGAR, BANGALORE, INDIA // PHOTOS © VIVEK MUTHURAMALINGAM

Architects: Biome Environmental Solutions
Design Team: Sharath Nayak, Chitra Vishwanath
Structural Engineer: Ravindranath B.V
Contractor: Narayan M.C.
Building Area: 140 m^2
Project Year: 2010

A dense urban context and a small site of 140 m^2 required a tall building with individual multipurpose studio spaces, one above the other, for a brother and sister to live and work. Building with earth was the preferred choice, using rammed earth instead of bricks. A basement space was not feasible for sourcing soil, so earth was sourced from a neighboring site. The two units are identical, each with a double height living space, a kitchenette, a powder room, and an L-shaped mezzanine space. The house is a box with strategic openings and a mix of frosted and clear glass. The double upper mezzanine is a revelation. This is a zero waste project that repurposed the formwork used for the rammed earth as the structure for the upper unit's mezzanine.

Ein dichtes städtisches Umfeld und eine kleine Fläche von 140 m^2 erforderten ein hohes Gebäude mit individuellen Mehrzweckstudios, die übereinander liegen, für einen Bruder und eine Schwester, um darin zu leben und zu arbeiten. Das Bauen mit Erde war die bevorzugte Wahl, wobei Stampflehm anstelle von Ziegeln verwendet wurde. Ein Kellerraum war nicht geeignet, um Boden für den Bau zu beschaffen, daher wurde die Erde von einem benachbarten Gelände bezogen. Die beiden Einheiten sind identisch, jede mit einem doppelt hohen Wohnraum, einer Küchenzeile, einem Gäste-WC und einem L-förmigen Zwischengeschoss. Das Haus ist ein Kasten mit strategisch platzierten Öffnungen und einer Mischung aus satiniertem und klarem Glas. Das doppelte obere Zwischengeschoss ist eine Offenbarung. Dies ist ein Null-Abfall-Projekt, das die Schalung, die für den Stampflehm verwendet wurde, als Struktur für das Zwischengeschoss der oberen Einheit wiederverwendet hat.

Un contexte urbain dense et un petit site de 140 m^2 nécessitaient un bâtiment haut avec des espaces de studio individuels polyvalents, l'un au-dessus de l'autre, pour qu'un frère et une sœur puissent vivre et travailler. Construire avec de la terre était le choix préféré, en utilisant de la terre battue au lieu de briques. Un espace de sous-sol n'était pas réalisable pour obtenir de la terre, donc la terre a été sourcée à partir d'un site voisin. Les deux unités sont identiques, chacune avec un espace de vie à double hauteur, une kitchenette, une salle de poudre et un espace mezzanine en forme de L. La maison est une boîte avec des ouvertures stratégiques et un mélange de verre dépoli et transparent. Le double mezzanine supérieur est une révélation. Il s'agit d'un projet zéro déchet qui a réutilisé le coffrage utilisé pour la terre battue comme structure pour le mezzanine supérieur de l'unité.

Un contexto urbano denso y un sitio pequeño de 140 m^2 requerían un edificio alto con espacios individuales de estudio multi-uso, uno encima del otro, para que un hermano y una hermana pudieran vivir y trabajar. La construcción con tierra fue la opción preferida, utilizando tierra apisonada en lugar de ladrillos. Un espacio de sótano no era factible para obtener tierra, por lo que se obtuvo de un sitio vecino. Las dos unidades son idénticas, cada una con un espacio de vida de doble altura, una cocina, un baño y un espacio tipo mezzanine en forma de L. La casa es una caja con aberturas estratégicas y una mezcla de vidrio esmerilado y transparente. El doble mezzanine superior es una revelación. Este es un proyecto de residuo cero que reutilizó el encofrado utilizado para la tierra apisonada como la estructura del mezzanine superior de la unidad.

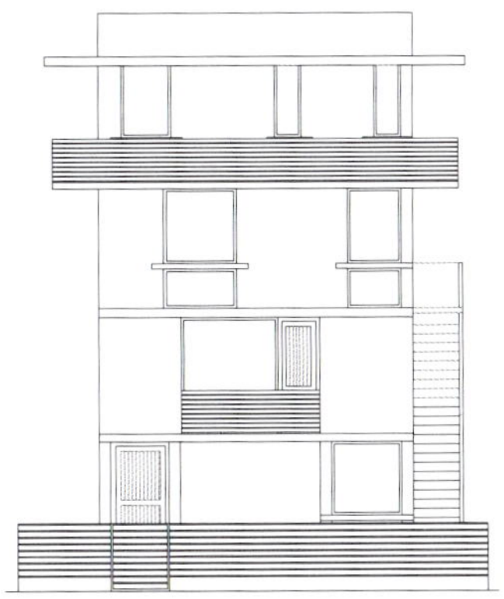

North elevation

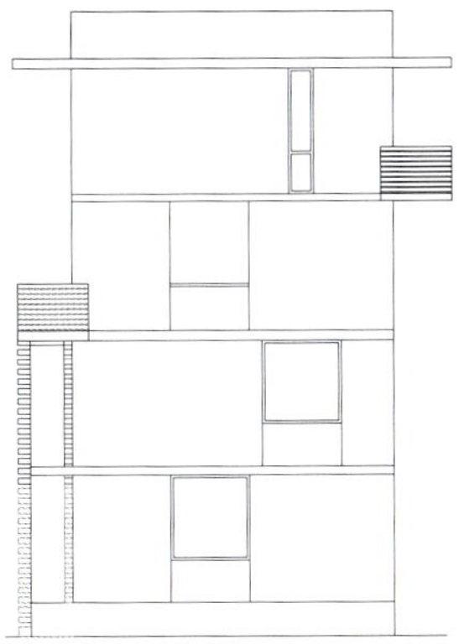

South elevation

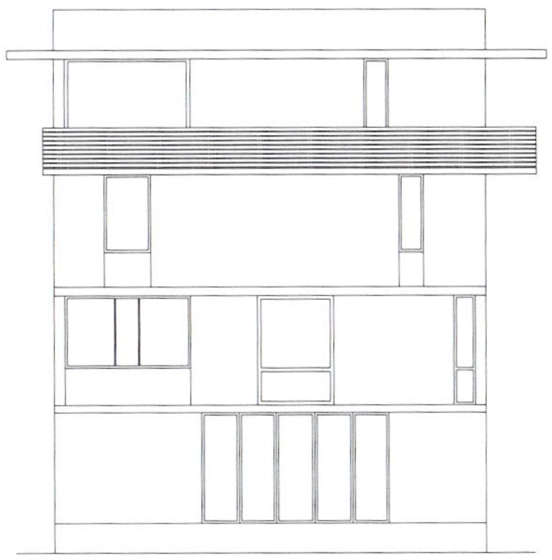

East elevation

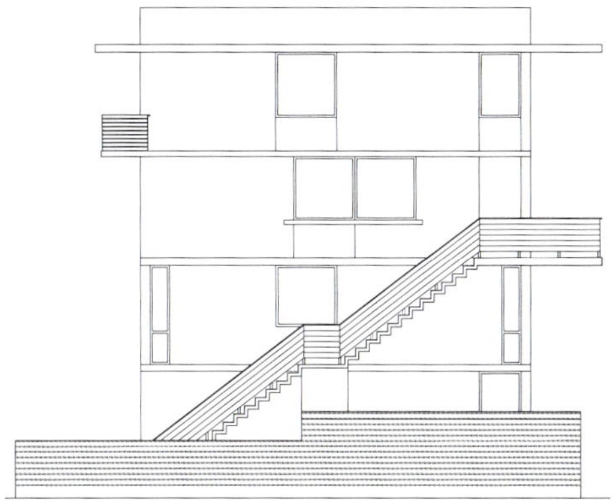

West elevation

0 1 2m

WAGHOBA ECOLODGE

CHANDRAPUR, MAHARASHTRA, INDIA // PHOTOS © LALIT RAJORIA

Architects: Biome Environmental Solutions
Design Team: Anurag Tamhankar, Chitra Vishwanath, Maitri Shah
Client: Pugdundee Safaris
Site Area: 3085.91 m^2
Project Year: 2021
Manufacturers: Hindware, Bathline, Adish Aluminium, The Purple Turtle
Consultants: Satish Raipure (Structural), HM Nagesh (PHE), NEERI (Water treatment system)
Contractors: Vikesh Agarwal
Furniture and Carpentry Work: SAGA (Doors), Adish Aluminium (Windows)
Conical Tile Vault Roofing: CSV Wardha
Swimming Pool: Nandadeep pools

Waghoba Ecolodge is a 14-cottage resort located in central India, adjacent to Tadoba wildlife sanctuary. The resort is designed for wildlife enthusiasts and conservationists, with responsible tourism at its core. The developers aimed to restore cultivated land to a deciduous forest and enhance biodiversity. A lake was created to provide water for the semi-arid climate and hold treated wastewater. Cottages use passive strategies to reduce cooling loads, and staff housing has a similar design to guest rooms. Clay sourced locally from desilted lakes is used throughout, as sun dried blocks for the walls and in the form of handmade terracotta tiles for the roofs. Sewage is treated with a plant-based system and used for growing organic vegetables and forest. Vaulted roofs were done in collaboration with CSV Wardha. The design approach is thus an intersection of ecology, social development, and leisure.

Waghoba Ecolodge ist ein 14-Hütten-Resort in Zentralindien, neben dem Tadoba-Wildtierreservat gelegen. Das Resort ist für Wildtierliebhaber und Naturschützer konzipiert und legt großen Wert auf verantwortungsvollen Tourismus. Die Entwickler hatten das Ziel, kultiviertes Land in einen Laubwald umzuwandeln und die Artenvielfalt zu fördern. Ein See wurde angelegt, um Wasser für das halbaride Klima bereitzustellen und behandeltes Abwasser aufzufangen. Die Hütten nutzen passive Strategien, um Kühllasten zu reduzieren, und Mitarbeiterunterkünfte haben ein ähnliches Design wie Gästezimmer. Ton, der aus örtlich entwässerten Seen gewonnen wird, wird im ganzen Resort verwendet, als sonnengetrocknete Blöcke für die Wände und als handgefertigte Terrakotta-Fliesen für die Dächer. Abwasser wird mit einer pflanzenbasierten Anlage behandelt und für den Anbau von Bio-Gemüse und für den Wald genutzt. Gewölbte Dächer wurden in Zusammenarbeit mit CSV Wardha ausgeführt. Der Designansatz ist somit eine Schnittstelle zwischen Ökologie, sozialer Entwicklung und Freizeit.

Waghoba Ecolodge est un complexe touristique de 14 chalets situé dans le centre de l'Inde, à côté du sanctuaire de la faune de Tadoba. Le complexe est conçu pour les passionnés de la faune et les conservationnistes, avec un tourisme responsable au cœur de ses préoccupations. Les développeurs ont cherché à restaurer des terres cultivées en une forêt caduque et à améliorer la biodiversité. Un lac a été créé pour fournir de l'eau pour le climat semi-aride et retenir les eaux usées traitées. Les chalets utilisent des stratégies passives pour réduire les charges de refroidissement, et le logement du personnel a un design similaire à celui des chambres d'hôtes. L'argile provenant localement des lacs désiltés est utilisée partout, sous forme de blocs séchés au soleil pour les murs et sous forme de tuiles de terracotta faites à la main pour les toits. Les eaux usées sont traitées avec un système à base de plantes et utilisées pour cultiver des légumes biologiques et des forêts. Les toits voûtés ont été réalisés en collaboration avec CSV Wardha. L'approche de conception est donc une intersection d'écologie, de développement social et de loisirs.

Waghoba Ecolodge es un complejo turístico de 14 cabañas ubicado en el centro de India, junto al santuario de vida silvestre de Tadoba. El complejo está diseñado para entusiastas de la vida silvestre y conservacionistas, con el turismo responsable en su núcleo. Los desarrolladores apuntaban a restaurar la tierra cultivada a un bosque caducifolio y mejorar la biodiversidad. Se creó un lago para proporcionar agua para el clima semiárido y retener las aguas residuales tratadas. Las cabañas utilizan estrategias pasivas para reducir las cargas de refrigeración, y el alojamiento del personal tiene un diseño similar al de las habitaciones de los huéspedes. La arcilla obtenida localmente de lagos desiltados se utiliza en todo el complejo, en forma de bloques secados al sol para las paredes y en forma de tejas de terracota hechas a mano para los techos. Las aguas residuales se tratan con un sistema a base de plantas y se utilizan para cultivar verduras orgánicas y bosques. Los techos abovedados se realizaron en colaboración con CSV Wardha. El enfoque de diseño es, por lo tanto, una intersección de ecología, desarrollo social y ocio.

Site plan

1. Watchman cabin
2. Guest facility area
3a. Guest rooms-north block
3b. Guest rooms-south block
4. Open well
5. Research facility
6. STP

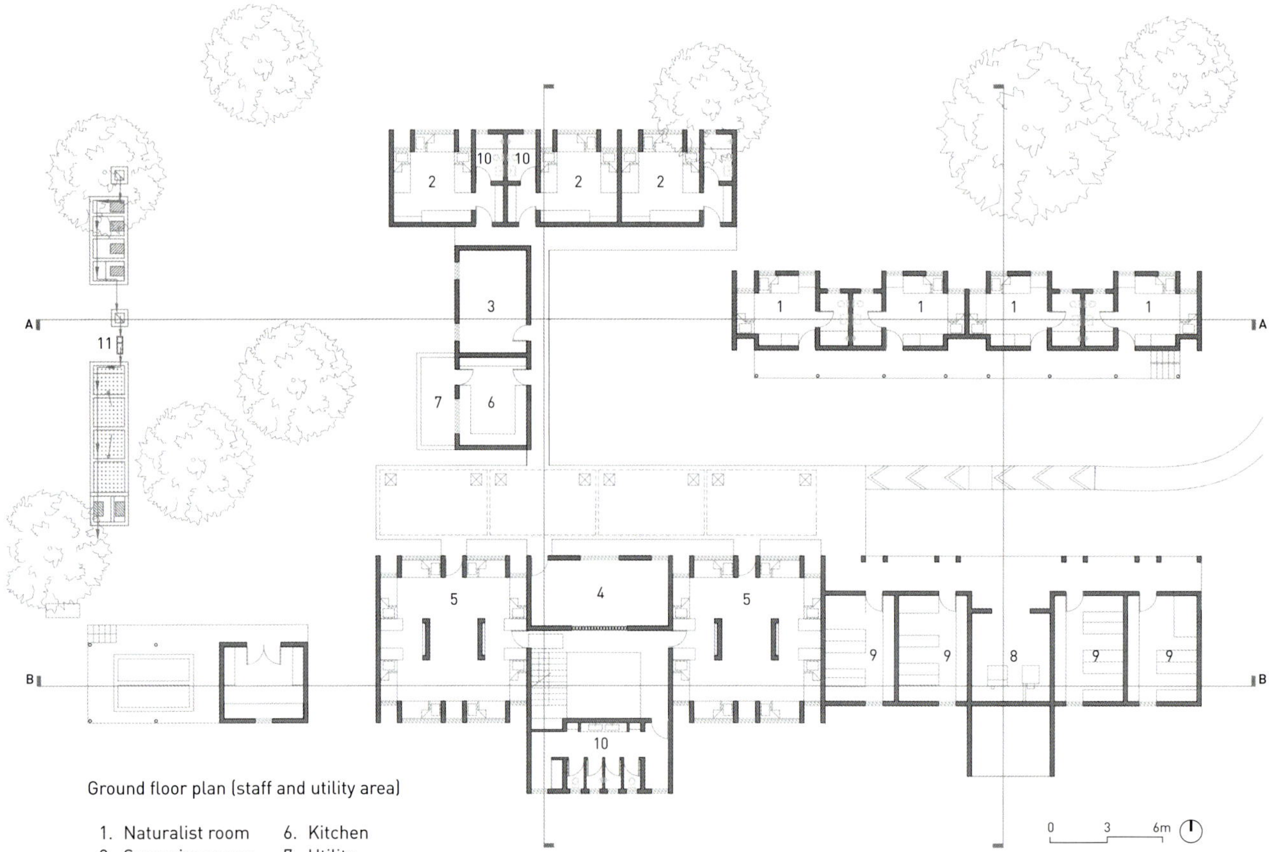

Ground floor plan (staff and utility area)

1. Naturalist room
2. Supervisor room
3. Executive dining area
4. Dining area
5. Dormitory
6. Kitchen
7. Utility
8. Laundry
9. Storage
10. Toilet
11. STP

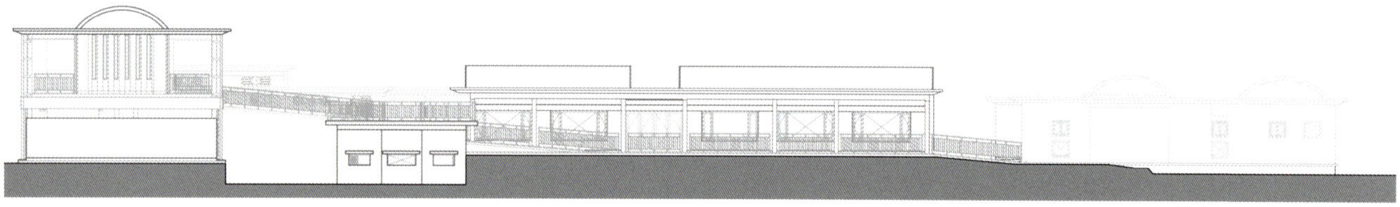

East elevation (guest facility area)

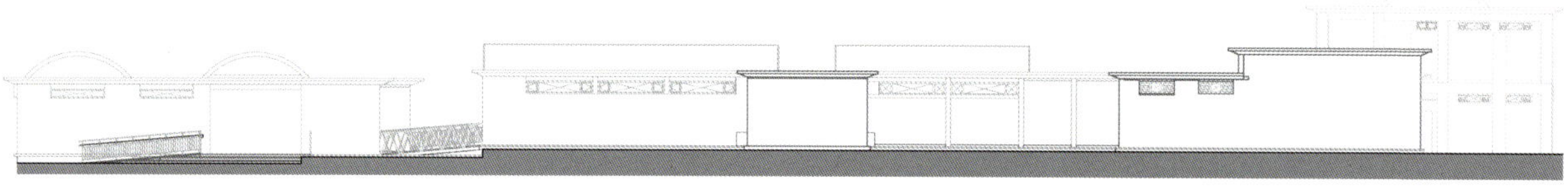

West elevation (guest facility area)

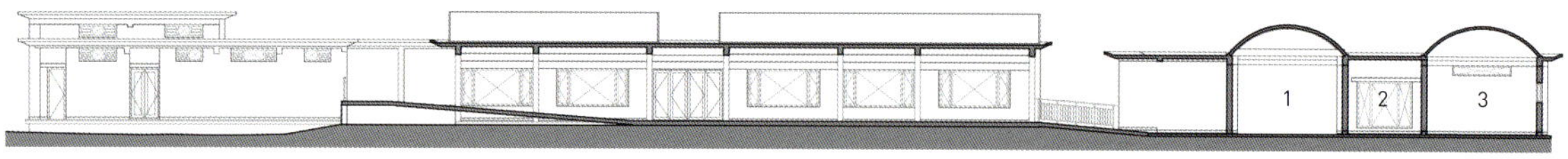

Section A-A (guest facility area)

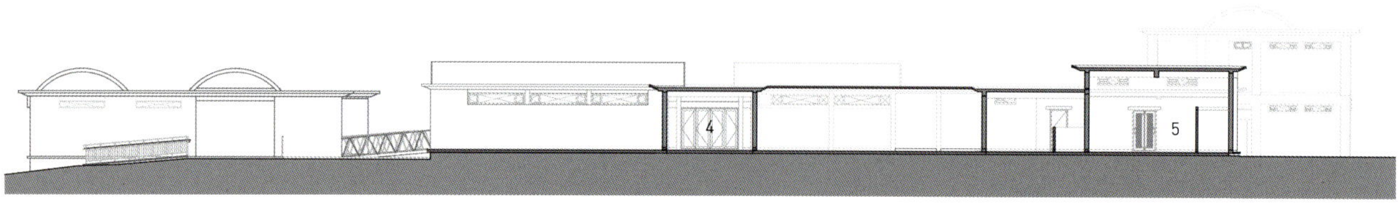

Section B-B (guest facility area)

1. Welcome area
2. Briefing area
3. Office area
4. Pantry
5. Kitchen

0 3 6m

CAMILO HOLGUIN - NATIVA

NATIVA is an architecture studio that combines local tradition with contemporary vision to design and build sustainable earth projects. Its focus on sustainability is inspired by the peasant notion of maximizing resources and minimizing losses. The studio uses bioclimatic strategies, permaculture, recycling of materials, and wind and solar energy, among others, to promote resilience over time of local knowledge and preserve building traditions. Additionally, it encourages the continuity of local trades in its projects and seeks to generate a positive impact on the lives of workers and on the natural and cultural landscape. Founded in 2003 by architect Camilo Holguín Ortiz, NATIVA has been recognized in the Colombian Architecture Biennial on several occasions and has participated in international events and workshops.

NATIVA ist ein Architekturstudio, das lokale Traditionen mit zeitgenössischer Vision kombiniert, um nachhaltige Erdbau-Projekte zu entwerfen und zu bauen. Der Fokus auf Nachhaltigkeit wird von der Bauernweisheit inspiriert, Ressourcen zu maximieren und Verluste zu minimieren. Das Studio nutzt bioklimatische Strategien, Permakultur, Recycling von Materialien, sowie Wind- und Solarenergie, um Resilienz im Laufe der Zeit zu fördern, das lokale Wissen zu bewahren und Bau-Traditionen zu erhalten. Darüber hinaus fördert es die Kontinuität lokaler Handwerksberufe in seinen Projekten und sucht nach positiven Auswirkungen auf das Leben der Arbeiter und auf die natürliche und kulturelle Landschaft. Gegründet wurde NATIVA 2003 vom Architekten Camilo Holguín Ortiz und wurde mehrmals in der Kolumbianischen Architekturbiennale ausgezeichnet und hat an internationalen Veranstaltungen und Workshops teilgenommen.

QUINTA DOS TUKANOS

NATIVA est un studio d'architecture qui combine la tradition locale avec une vision contemporaine pour concevoir et construire des projets durables avec la terre. Leur approche de la durabilité s'inspire de la notion paysanne de maximiser les ressources et de minimiser les pertes. Le studio utilise des stratégies bioclimatiques, de permaculture, de recyclage de matériaux et d'énergie éolienne et solaire, entre autres, pour promouvoir la résilience dans le temps des savoirs locaux et préserver les traditions constructives. En outre, il encourage la continuité des métiers locaux dans ses projets et cherche à avoir un impact positif sur la vie des travailleurs et sur le paysage naturel et culturel. Fondé en 2003 par l'architecte Camilo Holguín Ortiz, NATIVA a été reconnu à plusieurs reprises lors de la Biennale colombienne d'architecture et a participé à des événements et des ateliers internationaux.

NATIVA es un estudio de arquitectura que combina la tradición local con la visión contemporánea para diseñar y construir proyectos sustentables con tierra. Su enfoque en la sostenibilidad se inspira en la noción campesina de aprovechar al máximo los recursos y minimizar las pérdidas. El estudio utiliza estrategias bioclimáticas, permacultura, reciclaje de materiales y energía eólica y solar, entre otros, para promover la resiliencia en el tiempo de los saberes locales y preservar las tradiciones constructivas. Además, fomenta la continuidad de los oficios locales en sus proyectos y busca generar un impacto positivo en la vida de los trabajadores y en el paisaje natural y cultural. Fundado en 2003 por el arquitecto Camilo Holguín Ortiz, NATIVA ha sido reconocido en la Bienal Colombiana de Arquitectura en varias ocasiones y ha participado en eventos y talleres internacionales.

QUINTA DOS TUKANOS

PIRACAIA, SÃO PAULO, BRASIL // PHOTOS © CAMILO HOLGUÍN

Design Team: Arch. Daniel Parra,
Arch. Laura Angarita, Arch. Coral Fernandez,
Arch. Cesar Salgar, Arch. Camilo Holguín
Construction Team: Ing. Antonio Machado,
Arch. Camilo Holguín
Construction Period: 2018-2020
Height: 972 meters above sea level
Plot Area: 80.000 m^2
Built Area: 300 m^2
Terraces Area: 531,19 m^2
Construction Masters: Ciro Alfonso Mejia (COL),
Gerardo Gómez (COL), Juan Pablo Naranjo (COL),
Julio Alberto Sarmiento (COL), Dario Bohorquez (COL),
dom Freitas (Brasil)

The Quinta Dos Tukanos project, designed by a team of Colombo-Mexican architects, is an example of the transmission of knowledge of the tradition of building with earth from Barichara to another country, in this case, Brazil. The project consists of a new rammed earth and stone house, the renovation of an old structure on the property as an office, and the design of 4 cabins to host people who come to take permaculture workshops. The construction was carried out by a team of 5 master builders from Barichara, who traveled to Brazil to share their knowledge with the local workers. The entire construction process lasted approximately a year. The project stands out for its sustainability and bioclimatic design, using low-impact materials and efficient water management, with a wastewater treatment plant and solar systems for water heating.

Das Quinta Dos Tukanos-Projekt, das von einem Team kolumbianisch-mexikanischer Architekten entworfen wurde, ist ein Beispiel für die Übertragung des Wissens über die Tradition des Lehmbaus von Barichara in ein anderes Land, in diesem Fall Brasilien. Das Projekt umfasst ein neues Lehm- und Steinhaus, die Renovierung einer alten Struktur auf dem Grundstück als Büro und das Design von 4 Hütten, um Menschen zu beherbergen, die Permakultur-Workshops besuchen möchten. Die Bauarbeiten wurden von einem Team aus 5 Baumeistern aus Barichara durchgeführt, die nach Brasilien gereist sind, um ihr Wissen mit den örtlichen Arbeitern zu teilen. Der gesamte Bauvorgang dauerte etwa ein Jahr. Das Projekt zeichnet sich durch seine Nachhaltigkeit und bioklimatische Gestaltung aus und verwendet Materialien mit geringem Einfluss und eine effiziente Wasserwirtschaft mit einer Abwasserbehandlungsanlage und Solaranlagen zur Wassererwärmung.

Le projet Quinta Dos Tukanos, conçu par une équipe d'architectes colombo-mexicains, est un exemple de transmission des connaissances de la tradition de construction en terre de Barichara vers un autre pays, en l'occurrence le Brésil. Le projet comprend une nouvelle maison en pisé de terre et en pierre, la rénovation d'une ancienne structure de la ferme en tant que bureau, ainsi que la conception de 4 cabanes pour accueillir les personnes participant aux ateliers de permaculture. La construction a été confiée à une équipe de 5 maîtres d'œuvre de Barichara, qui sont venus au Brésil pour transmettre leurs connaissances à l'équipe de travailleurs locaux. Le processus de construction a duré environ un an. Le projet se distingue par sa durabilité et sa bioclimatique, en utilisant des matériaux à faible impact et une gestion efficace de l'eau, avec une usine de traitement des eaux usées et des systèmes solaires pour le chauffage de l'eau.

El proyecto Quinta Dos Tukanos, diseñado por un equipo de arquitectos Colombo-Mexicanos, es un ejemplo de la transmisión del conocimiento de la tradición de construcción con tierra de Barichara a otro país, en este caso, Brasil. El proyecto consta de una casa nueva de tapia pisada y piedra, la remodelación de una estructura antigua de la finca como oficina y el diseño de 4 cabañas para hospedar personas que vengan a tomar talleres de permacultura. La construcción estuvo a cargo de un equipo de 5 maestros de obra de Barichara, que viajaron a Brasil a transmitir sus conocimientos al equipo de trabajadores del lugar. Todo el proceso constructivo duró aproximadamente un año. El proyecto destaca por su sostenibilidad y bioclimática, usando materiales de bajo impacto y un manejo eficiente del agua, con una planta de tratamiento de aguas residuales y sistemas solares para el calentamiento del agua.

EL ALJIBE

BARICHARA, SANTANDER, COLOMBIA // PHOTOS © MATEO PÉREZ

Height: 1.336 meters above sea level
Medium Temperature: 23 °C
Plot Area: 790 m^2
Built Area: 384.87 m^2
Terraces Area: 153.99 m^2
Design: NATIVA – Arch. Camilo Holguín Ortiz
Construction: NATIVA - Arch. Camilo Holguín Ortiz
Construction Master: Feisar Roa
Upholsterer Masters: Alfonso, Roberto Garces
Landscape: Luis Alberto Martin
Collaborators: Luis Alberto Martin, Arch. José María Rodríguez, Arch. Juan Luis Isaza
Sketch: Andrés Rubio

Barichara, a Colombian town founded in 1705 that has preserved its material and immaterial heritage. Most of the buildings in Barichara are built with rammed earth, bahareque, adobe, stone, wood, and clay tiles, and the earth construction technique has been passed down from generation to generation. The place has attracted tourist attention, as well as people interested in living in the town and building their homes in a traditional way, leading to a renaissance of construction cultures. The house project in question is located on Calle Real in Barichara and had to comply with the regulations for the protection of the built heritage. The location of the lot on the high part of Barichara allowed for an impressive view of the town, and the construction of stepped terraces was necessary due to the topography of the land. In the conception of the project, the importance of an architecture respectful of the urban and architectural context of Barichara was taken into account.

Barichara, eine kolumbianische Stadt, die 1705 gegründet wurde und ihr materielles und immaterielles Erbe bewahrt hat. Die meisten Gebäude in Barichara sind aus Stampflehm, Bahareque, Adobe, Stein, Holz und Tonziegeln gebaut, und die Lehmbau-Technik wurde von Generation zu Generation weitergegeben. Der Ort hat touristische Aufmerksamkeit auf sich gezogen, sowie Menschen, die daran interessiert sind, in der Stadt zu leben und ihre Häuser auf traditionelle Weise zu bauen, was zu einer Wiedergeburt der Bautechniken führte. Das Hausprojekt in Frage befindet sich in der Calle Real in Barichara und musste den Vorschriften zum Schutz des gebauten Erbes entsprechen. Die Lage des Grundstücks auf dem hohen Teil von Barichara ermöglichte eine beeindruckende Aussicht auf die Stadt, und der Bau von gestuften Terrassen war aufgrund der Topographie des Geländes erforderlich. Bei der Konzeption des Projekts wurde die Bedeutung einer Architektur, die den städtischen und architektonischen Kontext von Barichara respektiert, berücksichtigt.

Barichara, un village colombien fondé en 1705 qui a préservé son patrimoine matériel et immatériel. La plupart des bâtiments de Barichara sont construits en pisé, en bahareque, en adobe, en pierre, en bois et en tuiles en terre cuite, et la technique de construction en terre a été transmise de génération en génération. Le lieu a attiré l'attention touristique, ainsi que des personnes intéressées à vivre dans la population et à construire leurs maisons de manière traditionnelle, ce qui a conduit à un renouveau des cultures constructives. Le projet de maison en question est situé dans la rue Real de Barichara et a dû respecter la réglementation de protection du patrimoine construit. L'emplacement du terrain sur la partie haute de Barichara a permis une vue impressionnante sur le village, et la construction de terrasses en escalier était nécessaire en raison de la topographie du terrain. Dans la conception du projet, l'importance d'une architecture respectueuse du contexte urbain et architectural de Barichara a été prise en compte.

Barichara, un pueblo colombiano fundado en 1705 que ha conservado su patrimonio material e inmaterial. La mayoría de las edificaciones en Barichara están construidas en tapia pisada, bahareque, adobe, piedra, madera y teja de barro, y se ha transmitido de generación en generación la técnica constructiva con tierra. El lugar ha atraído la atención turística, así como a personas interesadas en vivir en la población y construir sus casas de manera tradicional, lo que ha llevado a un renacimiento de las culturas constructivas. El proyecto de casa en cuestión se encuentra en la calle Real de Barichara y tuvo que cumplir con la normativa de protección del patrimonio construido. La ubicación del lote en la parte alta de Barichara permitió una vista impresionante del pueblo, y la construcción de terrazas escalonadas fue necesaria debido a la topografía del terreno. En la concepción del proyecto, se tuvo en cuenta la importancia de una arquitectura respetuosa del contexto urbano y arquitectónico de Barichara.

DI FRENNA ARQUITECTOS // MATIA DI FRENNA MÜLLER

We work linked to tradition and constantly experimenting with the materials and their techniques, with the support of the good and creative Mexican craftsmanship that we have. We understand architecture as a space-place, that its main purpose is to excite and at the same time fulfill the function for which it was designed. Functional spaces create our architecture, trying to match monumental and sculptural volumes with little maintenance, with this we create a very sober, functional and clean architecture.
"We are a young multidisciplinary architecture and design firm, with the ability to solve existing problems between users and their needs inside and outside the urban environment."

Wir arbeiten traditionsverbunden und experimentieren ständig mit Materialien und Techniken, unterstützt von guter und kreativer mexikanischer Handwerkskunst. Wir verstehen Architektur als einen Raum-Ort, dessen Hauptzweck es ist, zu begeistern und gleichzeitig die Funktion zu erfüllen, für die er entworfen wurde. Unsere Architektur besteht aus funktionalen Räumen, wobei wir versuchen, mit monumentalen und skulpturalen Volumina zu spielen, die nur wenig Wartung erfordern, wodurch wir eine sehr nüchterne, funktionale und saubere Architektur schaffen.
„Wir sind ein junges, multidisziplinäres Architektur- und Designbüro, das in der Lage ist, die bestehenden Probleme zwischen den Nutzern und ihren Bedürfnissen im und außerhalb des städtischen Umfelds zu lösen."

ESPACIO KAAB

Nous travaillons dans le respect de la tradition, en expérimentant constamment les matériaux et les techniques, avec le soutien de l'artisanat mexicain créatif et de qualité. Nous concevons l'architecture comme un espace-lieu, dont l'objectif principal est de susciter l'enthousiasme tout en remplissant la fonction pour laquelle il a été conçu. Les espaces fonctionnels créent notre architecture, en essayant de jouer avec des volumes monumentaux et sculpturaux avec peu d'entretien, ce qui nous permet de créer une architecture très sobre, fonctionnelle et propre.
« Nous sommes un jeune cabinet d'architecture et de design multidisciplinaire, capable de résoudre les problèmes existants entre les utilisateurs et leurs besoins dans et hors de l'environnement urbain ».

Trabajamos vinculados a la tradición, experimentando constantemente con los materiales y sus técnicas, con el apoyo de la buena y creativa mano de obra artesanal mexicana. Entendemos la arquitectura como espacio-lugar, que su destino principal es emocionar y que al mismo tiempo cumpla la función para la que fue diseñada. Los espacios funcionales crean nuestra arquitectura, intentando hacer juego de volúmenes monumentales y escultóricos con poco mantenimiento, con esto creamos una arquitectura muy sobria, funcional y limpia.
«Somos una firma joven multidisciplinaria de arquitectura y diseño, con la capacidad para resolver problemáticas existentes entre los usuarios y sus necesidades dentro y fuera del ámbito urbano».

ESPACIO KAAB

NOGUERAS, COMALA, COLIMA, MÉXICO // PHOTOS © LORENA DARQUEA

Finish Date: 15 december 2021
Area: 401.00 m²
Work Team: Arch. Matia Di Frenna Müller, Arch. Mariana de la Mora Padilla, Ing. Juan Gerardo Guardado Ávila
Project Type: Single Family Residential

Alejandro Rangel Hidalgo, known for his skill in the arts and crafts of Colima, lived in the Nogueras estate for many years. The Espacio Kaab project honors this site and the artisan's craft. The area is a unique ecosystem, with dry vegetation, cacti, stones, and earth. The house adapts to the terrain with a modular architectural program while respecting the topography. The main concept is the integration of indoor and outdoor spaces, using materials originating from Comala such as rammed earth walls and palm trees. The house has retaining walls that provide nooks and contemplation spaces. The intention is to generate open spaces with perspectives from inside the building. The interior design uses stone and wood furniture in its purest state, with indirect and highlighted lighting. Espacio Kaab is the latest addition to the Nogueras estate.

Alejandro Rangel Hidalgo, bekannt für seine Fähigkeiten in der Kunst und Handwerkskunst von Colima, lebte viele Jahre auf der Hacienda de Nogueras. Das Projekt Espacio Kaab ehrt dieses Anwesen und das Handwerk des Kunsthandwerkers. Die Region ist ein einzigartiges Ökosystem mit trockener Vegetation, Wüste, Kakteen, Steinen und Erde. Das Haus passt sich mit einem modularen architektonischen Programm an das Gelände an und respektiert die Topografie. Das Hauptkonzept ist die Integration von Innen- und Außenräumen unter Verwendung von Materialien aus Comala wie Lehm- und Palmwänden. Das Haus verfügt über Stützmauern, die Nischen und Betrachtungsräume schaffen. Die Absicht ist es, offene Räume mit Perspektiven von innen zu schaffen. Das Innendesign verwendet Möbel aus Stein und Holz in ihrem reinsten Zustand mit indirekter und hervorgehobener Beleuchtung. Espacio Kaab ist die neueste Ergänzung zur Hacienda de Nogueras.

Alejandro Rangel Hidalgo, connu pour son habileté dans les arts et l'artisanat de Colima, a vécu dans la propriété de Nogueras pendant de nombreuses années. Le projet Espacio Kaab rend hommage à ce lieu et à l'artisanat de l'artisan. La région est un écosystème unique, avec une végétation sèche, des cactus, des pierres et de la terre. La maison s'adapte au terrain avec un programme architectural modulaire tout en respectant la topographie. Le concept principal est l'intégration des espaces intérieurs et extérieurs, en utilisant des matériaux originaires de Comala tels que des murs en pisé et des palmiers. La maison a des murs de soutènement qui fournissent des recoins et des espaces de contemplation. L'intention est de générer des espaces ouverts avec des perspectives de l'intérieur du bâtiment. La conception intérieure utilise des meubles en pierre et en bois dans leur état le plus pur, avec un éclairage indirect et mis en valeur. Espacio Kaab est la dernière addition à la propriété de Nogueras.

Alejandro Rangel Hidalgo, conocido por su habilidad en el arte y artesanías de Colima, vivió en la Hacienda de Nogueras por muchos años. El proyecto Espacio Kaab honra este recinto y el oficio del artesano. La zona es un ecosistema único, con vegetación seca, árida, cactus, piedras y tierra. La casa se adapta al terreno con un programa arquitectónico modular, respetando la topografía. El concepto principal es la integración de los espacios interiores y exteriores, utilizando materiales originarios de Comala como muros de tapial y palma. La casa cuenta con muros de contención que dotan de recovecos y espacios de contemplación. La intención es generar espacios abiertos con perspectivas desde el interior de la construcción. El diseño interior utiliza muebles de piedra y madera en su estado más puro, con iluminación indirecta y destacada. Espacio Kaab es la más reciente adición a la hacienda de Nogueras.

Front elevation

Lateral elevation

Rear elevation

Lateral elevation

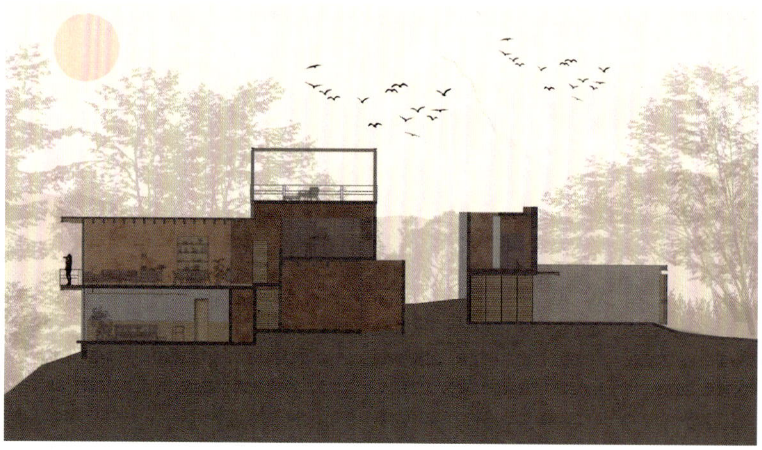

Sections

EDRA ARQUITECTURA KM0 // ÀNGELS CASTELLARNAU VISÚS

Edra Architecture km0 is an architecture studio located in Ayerbe, a small town in the province of Huesca in northern Spain. Edra works with low environmental impact materials, mainly earth and local fibers. The studio investigates traditional techniques and the availability of local material resources to develop new technologies that respond to the need for habitat and the climate emergency. Edra's architecture is contemporary, healthy, energy-efficient, and decarbonized, using passive techniques to bioclimatically resolve spaces while adapting to the natural and built environment in a calm and sustainable way.

Edra Architektur km0 ist ein Architekturstudio in Ayerbe, einem kleinen Dorf in der Provinz Huesca im Norden Spaniens. Edra arbeitet hauptsächlich mit umweltfreundlichen Materialien wie Erde und lokalen Fasern. Das Studio untersucht traditionelle Techniken und die Verfügbarkeit lokaler Materialressourcen, um neue Technologien zu entwickeln, die auf die Bedürfnisse des Lebensraums und des Klimanotfalls reagieren. Edras Architektur ist zeitgenössisch, gesund, energieeffizient und dekarbonisiert. Es verwendet passive Techniken, um Räume bioklimatisch zu lösen und sich dabei ruhig und nachhaltig an die natürliche und gebaute Umgebung anzupassen.

CASA DE TAPIA EN AYERBE

Edra Architecture km0 est un cabinet d'architecture situé à Ayerbe, petit village de la province de Huesca, au nord de l'Espagne. Edra travaille avec des matériaux à faible impact environnemental, principalement la terre et les fibres locales. Il étudie les techniques traditionnelles et la disponibilité des ressources matérielles locales pour développer de nouvelles technologies répondant aux besoins en habitat et à l'urgence climatique. L'architecture d'Edra est une architecture contemporaine, saine, énergétiquement efficace et décarbonisée, qui développe des techniques passives pour résoudre bioclimatiquement les espaces tout en s'adaptant à l'environnement naturel et construit de manière tranquille et en cherchant à rester.

Edra Arquitectura km0, es un estudio de arquitectura ubicado en Ayerbe, un pequeño pueblo de la provincia de Huesca al norte de España. Edra trabaja con materiales de bajo impacto ambiental principalmente con tierra y con fibras locales. Investiga las técnicas tradicionales y la disponibilidad de los recursos materiales locales para desarrollar nuevas tecnologías que den respuesta a la necesidad de hábitat y a la emergencia climática. La arquitectura de edra es una arquitectura contemporánea, saludable, energéticamente eficiente y descarbonizada, que desarrolla técnicas pasivas para resolver bioclimáticamente los espacios a la vez que se adapta al entorno natural y construido de forma tranquila y buscando permanecer.

HOTEL TORRE DE MARQUÉS

CASA DE TAPIA EN AYERBE

AYERBE, HUESCA, SPAIN // PHOTOS © XAVIER D'ARQUER DOBLESTUDIO

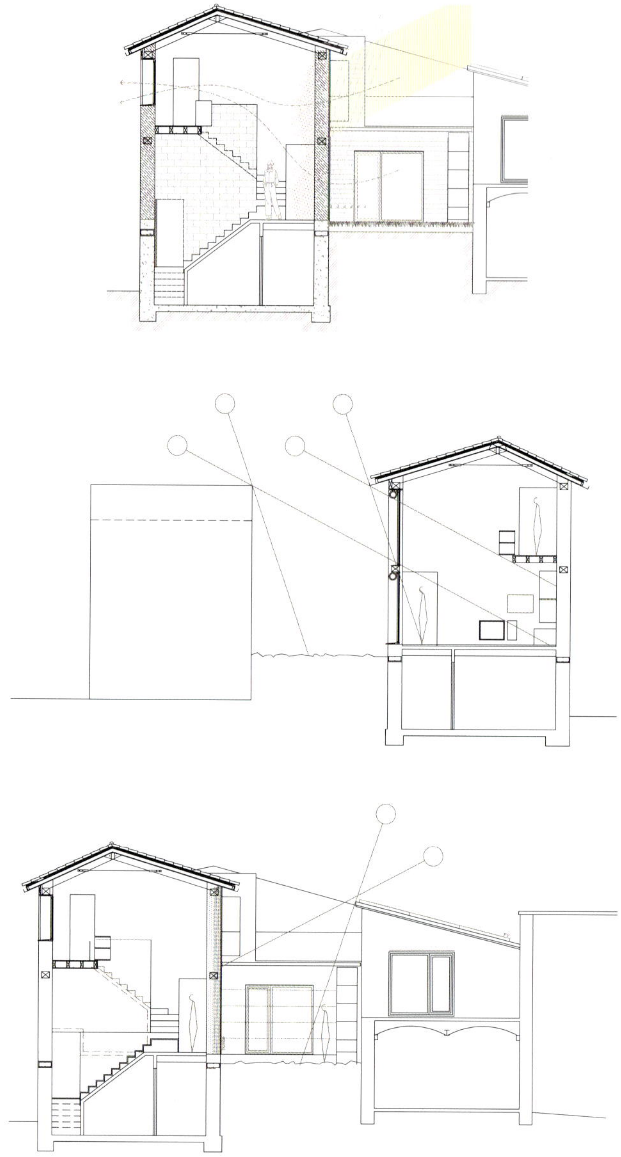

Sections

AWARDS:

- 1st Prize in the Housing category of the Terra Fibra Award 2016
- Sustainability Award from the García Mercadal Award in the housing category, given by the College of Architects of Aragon (2017)
- 1st Prize in the Housing category of the Castilla y León Sustainability Awards (2016)

Client: Alejandro Ascaso Sarasa
Quantity Surveyors: Fernando de Marcos

This project involves the construction of a house in a rural area of Ayerbe, in the Huesca region, in an area that has suffered depopulation and the loss of traditional use of land and building. The project focuses on the recovery of local identity and is committed to the environment and society. The design of the house is based on the vernacular architecture of the area and uses local materials such as earth, sandstone, wood, and reed. The construction technique used is "calicostrada tapia," which consists of building load-bearing walls with non-manufactured earth to which technical improvements are added. The house is divided into three floors, with the living spaces on the ground floor and an open patio facing south that allows for passive solar capture. The house has elements of protection against the sun and natural cross-ventilation. A dynamic thermal behavior strategy is used based on the inertia of the elements to achieve interior comfort. In addition, a trombe wall is being built on the south façade and its thermal behavior is being monitored in different orientations.

Dieses Projekt beinhaltet den Bau eines Hauses in einer ländlichen Gegend von Ayerbe, in der Region Huesca, in einem Gebiet, das unter Entvölkerung und dem Verlust der traditionellen Nutzung von Land und Gebäuden gelitten hat. Das Projekt konzentriert sich auf die Wiederherstellung der lokalen Identität und engagiert sich für Umwelt und Gesellschaft. Das Design des Hauses basiert auf der Volksarchitektur der Region und verwendet lokale Materialien wie Erde, Sandstein, Holz und Rohr. Die verwendete Bautechnik ist die „calicostrada tapia," bei der Tragwände mit nicht hergestellter Erde gebaut werden, zu der technische Verbesserungen hinzugefügt werden. Das Haus ist in drei Stockwerke unterteilt, mit den Wohnräumen im Erdgeschoss und einem offenen Patio nach Süden, der eine passive Sonnenenergieaufnahme ermöglicht. Das Haus verfügt über Elemente zum Schutz vor der Sonne und natürlicher Querlüftung. Eine Strategie für dynamisches thermisches Verhalten wird auf Basis der Trägheit der Elemente verwendet, um einen Innenkomfort zu erreichen. Darüber hinaus wird eine Trombe-Wand an der Südfassade gebaut und ihr thermisches Verhalten wird in verschiedenen Ausrichtungen überwacht.

Ce projet implique la construction d'une maison dans une zone rurale d'Ayerbe, dans la région de Huesca, dans une zone qui a subi la dépopulation et la perte de l'utilisation traditionnelle du sol et de la construction. Le projet vise à récupérer l'identité locale et s'engage en faveur de l'environnement et de la société. La conception de la maison est basée sur l'architecture vernaculaire de la région et utilise des matériaux locaux tels que la terre, la pierre de grès, le bois et le roseau. La technique de construction utilisée est la « tapia calicostrada », qui consiste en la construction de murs porteurs en terre non manufacturée à laquelle on ajoute des améliorations techniques. La maison est divisée en trois étages, avec des espaces de vie au rez-de-chaussée et une cour ouverte vers le sud qui permet une capture solaire passive. La maison dispose d'éléments de protection contre le soleil et la ventilation naturelle croisée. Une stratégie de comportement thermique dynamique est utilisée, basée sur l'inertie des éléments pour assurer le confort intérieur. De plus, un mur trombe est en train d'être construit sur la façade sud et son comportement thermique est surveillé dans différentes orientations.

Este proyecto implica la construcción de una casa en una zona rural de Ayerbe, en la región de Huesca, en un área que ha sufrido la despoblación y la pérdida del uso tradicional del suelo y la edificación. El proyecto se enfoca en la recuperación de la identidad local y está comprometido con el medio ambiente y la sociedad. El diseño de la casa se basa en la arquitectura vernácula de la zona y utiliza materiales locales como la tierra, la piedra arenisca, la madera y la caña. La técnica constructiva utilizada es la «tapia calicostrada», que consiste en la construcción de muros de carga con tierra no manufacturada a los que se añaden mejoras técnicas. La casa se divide en tres plantas, con las estancias vivideras en la planta baja y un patio abierto hacia el sur que permite la captación solar pasiva. La casa cuenta con elementos de protección contra el sol y la ventilación natural cruzada. Se utiliza una estrategia de comportamiento térmico dinámico basado en la inercia de los elementos para conseguir confort interior. Además, cuenta con un muro trombe en la fachada sur y se está monitorizando su comportamiento térmico en las distintas orientaciones.

North elevation

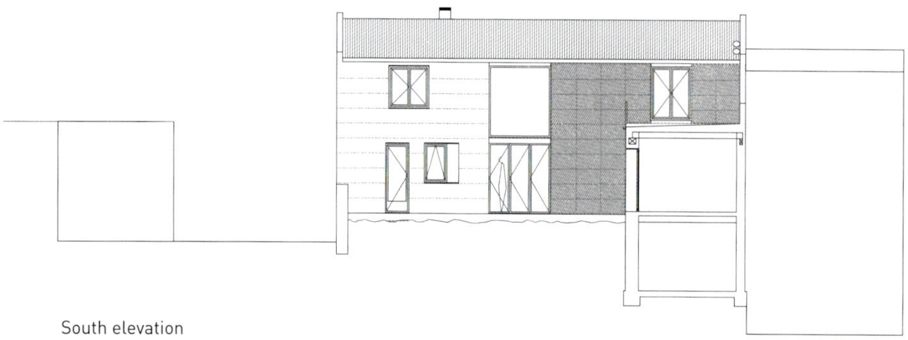

South elevation

HOTEL TORRE DEL MARQUÉS

TERUEL, ARAGON, SPAIN // PHOTOS © XAVIER D'ARQUER DOBLESTUDIO

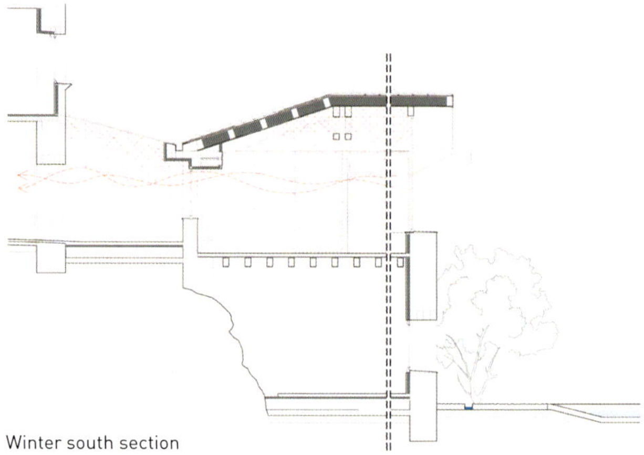

Winter south section

AWARDS:

- Selected in the Equipment category of Terra Fibra Award 2021
- Terra Ibérica 2021 Award organized by the College of Architects of Palencia and the Palencia Diputation
- Sustainability Award from the García Mercadal Award in the equipment category, given by the College of Architects of Aragon (2021)
- Best Rehabilitation Award in Aragon 2021 given by the College of Architects of Quantity Surveyors of Aragon

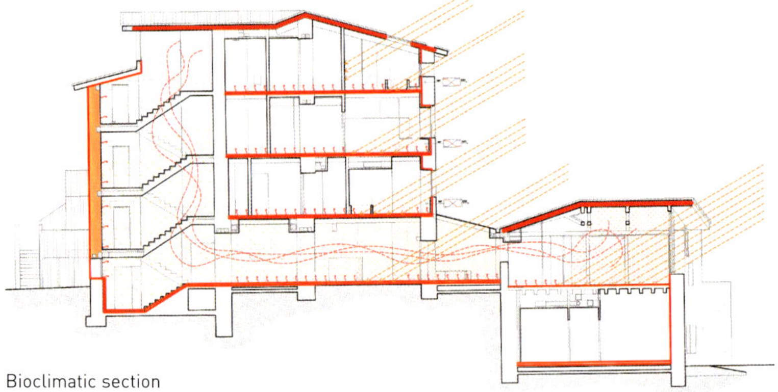

Bioclimatic section

Client: Inversiones del Matarranya SA
Engineering: Fernando Calavera Opi
Quantity Surveyors: Fernando de Marcos Calvo, Jose Miguel Sanz Lahoz
Interior Design: Sara Fernandez España
Landscaping: Pepa Morán

In Catalonia and Aragon, masias are autonomous and sustainable traditional agricultural farms. The rehabilitation and expansion project of the Marques Tower in Monroyo, Teruel, turned a 17th-century masia into a four-star hotel with decarbonization strategies. 90 % local materials and traditional construction techniques were used, resulting in an extremely effective circular economy project with an extremely low ecological footprint. Energy demand was reduced through careful bioclimatic strategy and self-generation of energy. In agronomic terms, crops were reactivated and sustainable forest management was initiated to produce chips. The building was developed respecting local construction culture and promoting innovation and reduction of environmental impact. The materials used to create the spaces included clay, plaster, stone, wood, straw, hemp, and light, creating healthy spaces free from toxic additives. Additionally, rainwater was collected and recirculated for irrigation.

In Katalonien und Aragonien gibt es autonome Masias, die traditionelle, nachhaltige landwirtschaftliche Betriebe sind. Der Turm von Marqués in Monroyo, Teruel, wurde durch ein Rehabilitations- und Erweiterungsprojekt zu einem Vier-Sterne-Hotel mit Dekarbonisierungsstrategien umgewandelt. Dabei wurden 90 % lokale Materialien und traditionelle Baumethoden verwendet, was zu einem effektiven Kreislaufprojekt mit geringem ökologischen Fußabdruck führte. Eine bioklimatische Strategie und Eigenenergieerzeugung reduzierten den Energiebedarf. Zudem wurden Kulturen reaktiviert und eine nachhaltige Forstwirtschaft zur Produktion von Hackschnitzeln gestartet. Das Gebäude wurde unter Berücksichtigung der lokalen Baukultur entwickelt, fördert Innovation und reduziert die Umweltauswirkungen. Materialien wie Ton, Gips, Stein, Holz, Stroh, Hanf und Licht wurden für die Räume verwendet, um gesunde Räume frei von giftigen Zusatzstoffen zu schaffen.

La réhabilitation de la Torre del Marqués à Monroyo, Teruel, a transformé une masia traditionnelle en un hôtel quatre étoiles avec une empreinte écologique extrêmement faible. 90 % des matériaux utilisés étaient locaux, les techniques de construction traditionnelles ont été utilisées et une gestion forestière durable a été lancée pour produire des éclats de bois. La demande énergétique a été réduite grâce à une stratégie bioclimatique et à l'autoproduction d'énergie. Les matériaux utilisés pour créer les espaces comprenaient de l'argile, du plâtre, de la pierre, du bois, de la paille, du chanvre et de la lumière, créant ainsi des espaces sains et exempts d'additifs toxiques. Le projet a encouragé l'innovation et la réduction de l'impact environnemental tout en respectant la culture constructive locale et en réactivant les cultures agricoles. C'est un exemple réussi d'économie circulaire et de durabilité dans l'agriculture et l'hôtellerie.

En Cataluña y Aragón, las masías son explotaciones agrícolas tradicionales autónomas y sostenibles. El proyecto de rehabilitación y ampliación de la Torre del Marqués en Monroyo, Teruel, convirtió una masía del siglo XVII en un hotel de cuatro estrellas con estrategias de descarbonización. Se utilizó un 90 % de materiales locales y técnicas de construcción tradicionales, lo que resultó en un proyecto de economía circular muy efectivo con una huella ecológica extremadamente baja. La demanda energética se redujo gracias a una estrategia bioclimática cuidadosa y la autogeneración de energía. En términos agronómicos, se reactivaron los cultivos y se inició una gestión forestal sostenible para producir astillas. El edificio se desarrolló respetando la cultura constructiva local y fomentando la innovación y la reducción del impacto ambiental. Los materiales utilizados para crear los espacios incluyeron arcilla, yeso, piedra, madera, paja, cáñamo y luz, creando espacios saludables y libres de aditivos tóxicos.

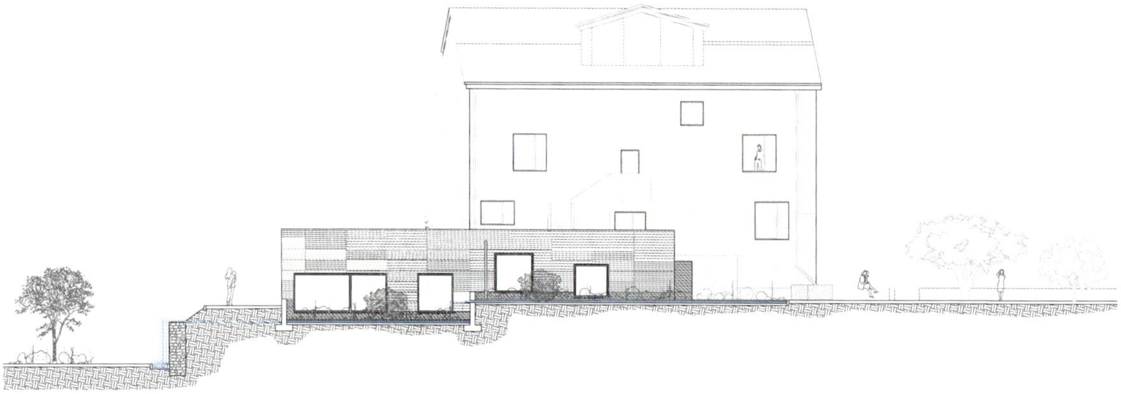

North elevation

South elevation

EMBAIXADA // CRISTINA MENDONÇA - NUNO GRIFF E PAULO GOINHAS

Embaixada was founded in 2002 by seven architects who aimed to produce innovative projects capable of meeting contemporary life's requirements. The office refuses to specialize in just one area and is constantly looking for new challenges. Their work includes Urbanism, Architecture, New Media, Arts, and Design. Unleash was established in 2012 to clarify the overlapping nature of the design and research phases, and it focuses on addressing processes and outputs on the architecture frontier with an academic approach. Embaixada and Unleash are two sides of the same coin, influenced by other cultural entities, but they hold the ambition to develop processes that refuse generalist results. They develop each project as a prototype, considering the identity of each design and the programmatic conditions of each client. The hybrid team relies on several specialists in different areas, which enables them to present a diversified and dynamic structure. They hold debates and avoid the traditional way of decision-making, generating a strong creative potential and an ample capacity of adaptation to various types of projects. Recently, Embaixada was invited to curate the next Lisbon Open House 2023.

Embaixada wurde 2002 von sieben Architekten gegründet, die innovative Projekte produzieren wollten, die den Anforderungen des zeitgenössischen Lebens gerecht werden. Das Büro lehnt es ab, sich nur auf einen Bereich zu spezialisieren, und ist ständig auf der Suche nach neuen Herausforderungen. Ihr Arbeitsbereich umfasst Stadtplanung, Architektur, Neue Medien, Kunst und Design. Unleash wurde 2012 gegründet, um die überlappende Natur der Design- und Forschungsphasen zu klären und konzentriert sich auf die Bearbeitung von Prozessen und Ergebnissen im Bereich Architektur mit einem wissenschaftlichen Ansatz. Embaixada und Unleash sind zwei Seiten derselben Medaille, beeinflusst von anderen kulturellen Einrichtungen, aber sie haben die Ambition, Prozesse zu entwickeln, die sich gegen generalistische Ergebnisse wehren. Jedes Projekt wird als Prototyp entwickelt, wobei die Identität jedes Designs und die programmatischen Bedingungen jedes Kunden berücksichtigt werden. Das Hybrid-Team verlässt sich auf mehrere Spezialisten in verschiedenen Bereichen, was es ihnen ermöglicht, eine vielfältige und dynamische Struktur zu präsentieren. Sie führen Debatten und vermeiden die traditionelle Entscheidungsfindung, was ein starkes kreatives Potenzial und eine große Anpassungsfähigkeit an verschiedene Projekte ermöglicht. Kürzlich wurde Embaixada eingeladen, das nächste Open House im Jahr 2023 zu kuratieren.

HERDADE DA CARDEIRA

Embaixada a été fondée en 2002 par sept architectes qui ont cherché à produire des projets innovants capables de répondre aux exigences de la vie contemporaine. Le bureau refuse de se spécialiser dans un seul domaine et est constamment à la recherche de nouveaux défis. Leurs travaux comprennent l'urbanisme, l'architecture, les nouveaux médias, les arts et le design. Unleash a été créé en 2012 pour clarifier la nature chevauchante des phases de conception et de recherche, et se concentre sur l'adresse des processus et des résultats à la frontière de l'architecture avec une approche académique. Embaixada et Unleash sont deux faces d'une même pièce, influencées par d'autres entités culturelles, mais ayant l'ambition de développer des processus qui refusent des résultats généralistes. Ils développent chaque projet comme un prototype, en tenant compte de l'identité de chaque design et des conditions programmatiques de chaque client. L'équipe hybride s'appuie sur plusieurs spécialistes dans différents domaines, ce qui leur permet de présenter une structure diversifiée et dynamique. Ils organisent des débats et évitent la méthode traditionnelle de prise de décision, générant un fort potentiel créatif et une grande capacité d'adaptation à divers types de projets. Récemment, Embaixada a été invitée à organiser la prochaine journée portes ouvertes en 2023.

Embaixada fue fundada en 2002 por siete arquitectos con el objetivo de producir proyectos innovadores capaces de satisfacer los requisitos de la vida contemporánea. La oficina se niega a especializarse en un solo campo y está constantemente buscando nuevos desafíos. Su trabajo incluye Urbanismo, Arquitectura, Nuevos Medios, Arte y Diseño. Unleash se estableció en 2012 para aclarar la naturaleza superpuesta de las fases de diseño e investigación, y se centra en abordar los procesos y resultados en la vanguardia de la arquitectura con un enfoque académico. Embaixada y Unleash son dos caras de la misma moneda, influenciadas por otras entidades culturales, pero que tienen la ambición de desarrollar procesos que rechacen resultados generalistas. Desarrollan cada proyecto como un prototipo, considerando la identidad de cada diseño y las condiciones programáticas de cada cliente. El equipo híbrido cuenta con varios especialistas en diferentes áreas, lo que les permite presentar una estructura diversificada y dinámica. Realizan debates y evitan el método tradicional de toma de decisiones, generando un gran potencial creativo y una amplia capacidad de adaptación a diversos tipos de proyectos. Recientemente, Embaixada fue invitada a curar la próxima jornada de puertas abiertas en 2023.

HERDADE DA CARDEIRA

ORADA, BORBA, PORTUGAL // PHOTOS © SIMONE BOSSI

AWARDS:

- 2021 "Herdade da Cardeira" Borba, "Architecture, Susteinability and Inovation award 2021" Lisboa, OA, Finalist
- 2012 "Agrob Buchtal Exhibition Stand," Cersaie fair, "Ait Global Award 2012," Winner
- 2009 "Casa dos Cubos" Tomar, Contract World Award, Conversion Category Winner
- 2009 "Casa dos Cubos" Tomar, Mies van der Rohe Award, Nominee
- 2008 "Embaixada Best Office," Prémios Construir, Nominee

Client: Herdade da Cardeira, Erika & Thomas Meier
Landscape Architect: Sandra Nunes
Civil Engineer: Pedro Fragoso Viegas
Rammed Earth Consulting: Henrique Schreck
Building Contractor: Construções Monsaraz

This project consisted of the extension of an existing winery at Herdade da Cardeira in Borba, Portugal. A house and office area were designed to support the winery's activities and the aim was to integrate the existing buildings with the new construction. The strategy involved flattening the terrain and placing the new buildings at different levels along the hill to accommodate the privacy needs of each use. Local materials and cross-ventilation techniques were used to improve energy efficiency. The project took eight years to complete and focused on seamless integration between indoor and outdoor spaces, and between residential and winery areas.The property is conceived as a single entity where agriculture, commerce and housing share space.

Dieses Projekt beinhaltete die Erweiterung einer bestehenden Kellerei in der Herdade da Cardeira in Borba, Portugal. Es wurde ein Haus und ein Bürobereich entworfen, um die Aktivitäten der Kellerei zu unterstützen, wobei die bestehenden Gebäude in die neue Struktur integriert wurden. Die Strategie bestand darin, das Gelände zu modellieren und die neuen Gebäude auf verschiedenen Ebenen entlang des Hügels zu platzieren, um den Bedarf an Privatsphäre für jeden Zweck zu erfüllen. Es wurden lokale Materialien und Techniken der Querlüftung eingesetzt, um die Energieeffizienz zu verbessern. Das Projekt dauerte acht Jahre und konzentrierte sich auf eine nahtlose Integration von Innen- und Außenräumen sowie zwischen Wohn- und Kellerbereichen. Das Anwesen wird als eine Einheit betrachtet, in der Landwirtschaft, Handel und Wohnen Raum teilen.

Ce projet consistait en l'extension d'une cave existante à Herdade da Cardeira à Borba, Portugal. Une maison et une zone de bureaux ont été conçues pour soutenir l'activité de la cave, en cherchant à intégrer les constructions existantes avec la nouvelle construction. La stratégie consistait à adoucir la topographie du terrain et à placer les nouvelles constructions à différents niveaux le long de la colline, afin de répondre aux besoins de confidentialité pour chaque utilisation. Des matériaux locaux et des techniques de ventilation croisée ont été utilisés pour améliorer l'efficacité énergétique. Le projet a pris huit ans pour être réalisé et s'est concentré sur une intégration fluide entre les espaces intérieurs et extérieurs, et entre les zones résidentielles et de la cave. La propriété est conçue comme une entité unique où l'agriculture, le commerce et le logement partagent l'espace.

Este proyecto consistió en la expansión de una bodega existente en la Herdade da Cardeira en Borba, Portugal. Se diseñó una casa y una zona de oficinas para apoyar la actividad de la bodega y se buscó integrar las construcciones preexistentes con la nueva edificación. La estrategia consistió en suavizar la topografía del terreno y ubicar las nuevas construcciones en diferentes niveles a lo largo de la colina, para adaptarse a las necesidades de privacidad de cada uso. Se utilizaron materiales locales y técnicas de ventilación cruzada para mejorar la eficiencia energética. El proyecto tomó ocho años para completarse y se enfocó en una integración fluida entre los espacios interiores y exteriores, y entre las áreas residenciales y de la bodega. La propiedad se concibe como una sola entidad donde la agricultura, el comercio y la vivienda comparten espacio.

Site plan

A. Entrance gate
B. Wine cellar (pre-existent)
C. Wine cellar support area
D. Housing area (social)
E. Housing (bedrooms)
F. Parking
G. Power and water supply house | recycling waste

1. Entrance path
2. Refreshment pergola
3. Olive grove terrace
4. Vine terrace
5. "Monte" hill
6. Vine path
7. Olive grove path
8. Pool terrace
9. White wisterias path

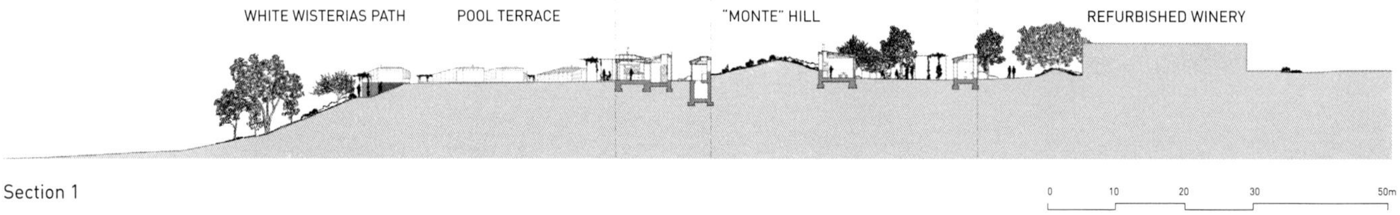

Section 1

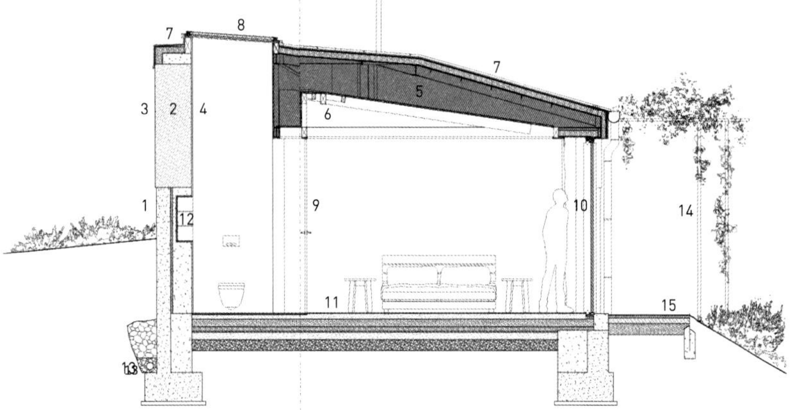

1. Structural concrete
2. Rammed earth
3. Lime plaster, refined to the colour of the rammed earth
4. Lime plaster, white colour
5. Light steel frame
6. Structure in solid pine wood
7. Covering system in natural zinc
8. Roof window in anodised aluminium, gray colour
9. Solid door in riga wood (reused)
10. Sliding door in anodised aluminium, light champagne colour
11. Flooring in polished concrete
12. Hidraulic mosaic
13. Drainage system
14. Pergola in steel, dark grey colour
15. Stabilised gravel with corten steel

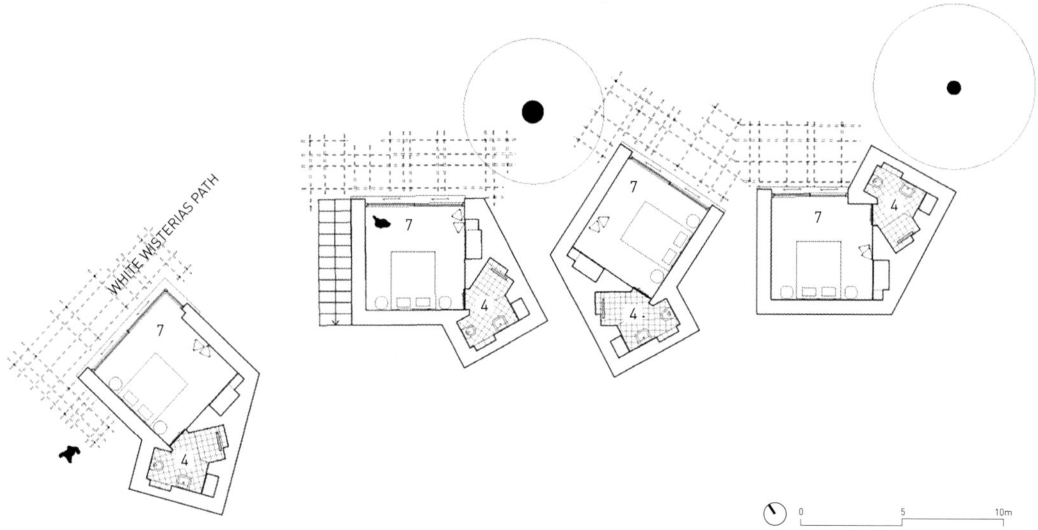

Housing area plan

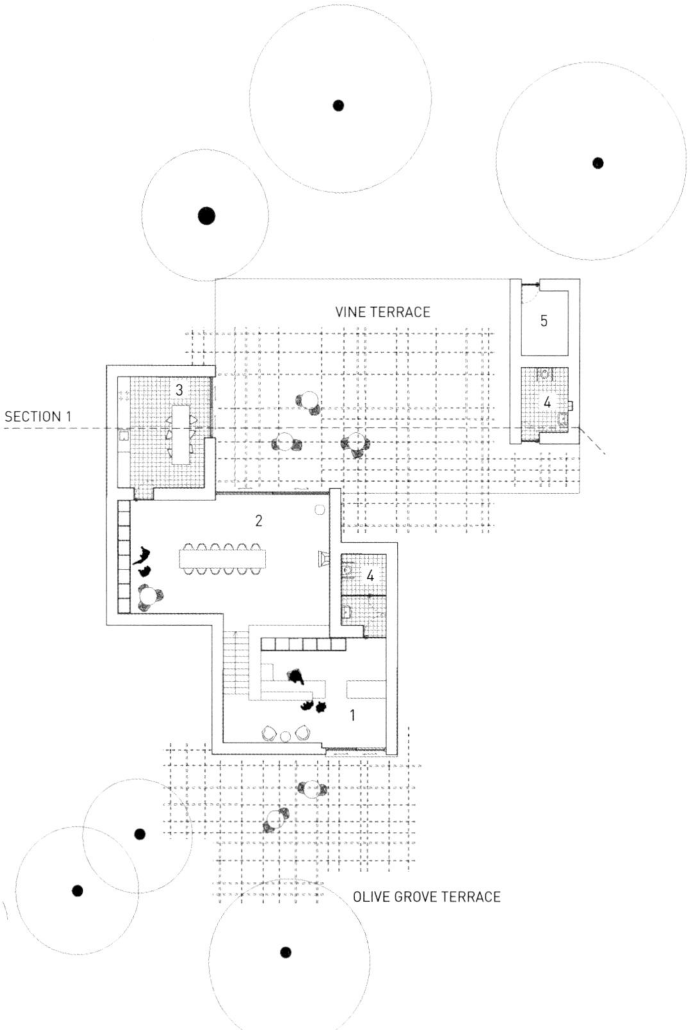

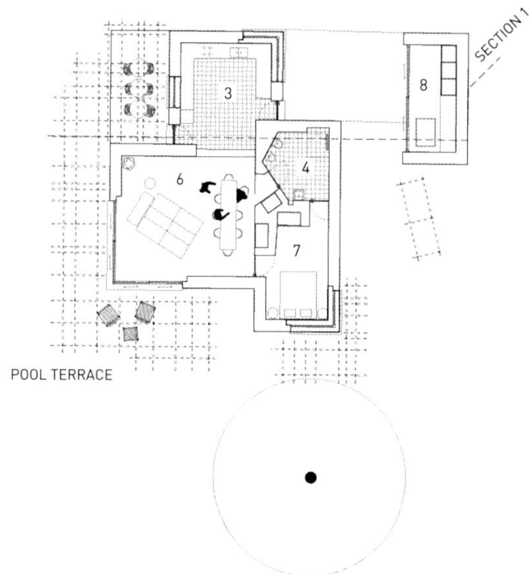

Housing area (social) plan

Wine cellar support area plan

1. Office | Reception
2. Tasting room
3. Kitchen
4. WC
5. Storage
6. Living room
7. Bedroom
8. Laundry

GLH ARCHITECTS // XAVIER HUYBERECHTS - DAMIEN HUYBERECHTS

GLH Architects (Grosskopff, Lombart, Huyberechts & Associates) is an architectural design firm based in Johannesburg. We design and build projects in South Africa and internationally. We have been in practice since 1945. Although we are one of the oldest architecture firms in continuous practice in the country, we have a young and dynamic team able to consistently produce good architecture; always maintaining design integrity and quality in process and product. We are a medium-sized firm of around 30 people. In maintaining a workforce of this size we balance the agility of a small studio with the resources of a larger firm, resulting in each building receiving the utmost consideration from concept to implementation. Our deep love of architecture and place-making drives our aim to maintain the highest levels of design innovation through the creation of exciting yet timeless buildings which graciously hold their own in the built environment. Our ethos is that our architecture adds value to any project.

GLH Architects (Grosskopff, Lombart, Huyberechts & Associates) ist ein Architekturbüro mit Sitz in Johannesburg. Wir entwerfen und bauen Projekte in Südafrika und international. Wir sind seit 1945 tätig. Obwohl wir eines der ältesten Architekturbüros in kontinuierlicher Praxis im Land sind, verfügen wir über ein junges und dynamisches Team, das kontinuierlich gute Architektur produzieren kann, wobei Designintegrität und Qualität im Prozess und im Produkt stets gewahrt bleiben. Wir sind ein mittelgroßes Unternehmen mit etwa 30 Mitarbeitern. Durch die Aufrechterhaltung einer Belegschaft dieser Größe balancieren wir die Agilität eines kleinen Studios mit den Ressourcen eines größeren Unternehmens, wodurch jedes Gebäude von der Konzeption bis zur Umsetzung die höchste Aufmerksamkeit erhält. Unsere tiefe Liebe zur Architektur und zum Platzmachen treibt unser Ziel an, die höchsten Maßstäbe an Designinnovation durch die Schaffung aufregender und gleichzeitig zeitloser Gebäude zu erhalten, die sich gnädig in der gebauten Umgebung behaupten. Unsere Ethik ist, dass unsere Architektur jedem Projekt einen Mehrwert verleiht.

WITKLIPFONTEIN ECO LODGE

GLH Architects (Grosskopff, Lombart, Huyberechts & Associates) est un cabinet de conception architecturale basé à Johannesburg. Nous concevons et construisons des projets en Afrique du Sud et à l'international. Nous sommes en activité depuis 1945. Bien que nous soyons l'une des plus anciennes entreprises d'architecture en pratique continue dans le pays, nous disposons d'une équipe jeune et dynamique capable de produire constamment une bonne architecture tout en maintenant l'intégrité et la qualité de la conception dans le processus et le produit. Nous sommes une entreprise de taille moyenne d'environ 30 personnes. En maintenant une main-d'œuvre de cette taille, nous équilibrons l'agilité d'un petit studio avec les ressources d'une entreprise plus importante, ce qui permet à chaque bâtiment de recevoir la plus grande attention de la conception à la mise en œuvre. Notre profonde passion pour l'architecture et la création de lieux guide notre objectif de maintenir les plus hauts niveaux d'innovation de conception en créant des bâtiments passionnants et intemporels qui s'intègrent gracieusement dans l'environnement bâti. Notre éthique est que notre architecture ajoute de la valeur à tout projet.

GLH Architects (Grosskopff, Lombart, Huyberechts & Associates) es una firma de diseño arquitectónico con sede en Johannesburgo. Diseñamos y construimos proyectos en Sudáfrica y a nivel internacional. Hemos estado en práctica desde 1945. Aunque somos una de las firmas de arquitectura más antiguas en práctica continua en el país, contamos con un equipo joven y dinámico capaz de producir consistentemente buena arquitectura; siempre manteniendo la integridad y calidad del diseño en el proceso y producto. Somos una firma de tamaño mediano de alrededor de 30 personas. Al mantener una fuerza laboral de este tamaño, equilibramos la agilidad de un pequeño estudio con los recursos de una firma más grande, lo que resulta en que cada edificio reciba la mayor consideración desde el concepto hasta la implementación. Nuestro profundo amor por la arquitectura y la creación de lugares impulsa nuestro objetivo de mantener los más altos niveles de innovación en el diseño a través de la creación de edificios emocionantes pero atemporales que mantienen su propia presencia en el entorno construido. Nuestro ethos es que nuestra arquitectura agrega valor a cualquier proyecto.

WITKLIPFONTEIN ECO LODGE

VREDEFORT, FREE STATE, SOUTH AFRICA // PHOTOS © ELSA YOUNG

Architects: Xavier Huyberechts & Damien Huyberechts, GLH Architects

Consultants:

- Quantity Surveying: Xavier Huyberechts, GLH Architects
- Project Manager: Damien Huyberechts
- Civil Engineering: Craig Thompson, PURE Engineers
- Mechanical Engineering: Xavier Huyberechts, GLH Architects
- Principal Contractor: Damien Huyberechts
- Electrical Engineering: Damien Huyberechts
- Landscape: Xavier Huyberechts, GLH Architects and Damien Huyberechts
- Green/Sustainable Consultants: Xavier Huyberechts, GLH Architects and Damien Huyberechts

Client: Xavier Huyberechts and Carine Huyberechts - Bwakira

Contractors: Damien Huyberechts, main contractor

Timeline: 2016-2018

Area: 400 m^2

The lodge is an architecturally unique building that seamlessly integrates into its natural surroundings through the use of traditional vernacular building techniques and modern minimalistic design. Its aim is to rethink the concept of a house/lodge by utilizing eco-friendly practices and technologies that are respectful to the environment. Overlooking the plains, the lodge uses the principle of thermal mass to flatten temperatures throughout the year, avoiding mechanical air conditioning. It also features a range of sustainable innovations, such as cement-free construction, recycled granite floors, and solar water heating. Overall, the lodge serves as a testament to the idea that traditional building techniques, such as rammed earth, can be reinterpreted in a modern way to create innovative and environmentally friendly architecture.

Die Lodge ist ein architektonisch einzigartiges Gebäude, das durch die Verwendung traditioneller Volkstechniken und moderner minimalistischer Gestaltung nahtlos in seine natürliche Umgebung integriert wird. Ihr Ziel ist es, das Konzept eines Hauses/Lodges durch die Nutzung umweltfreundlicher Praktiken und Technologien, die respektvoll gegenüber der Umwelt sind, neu zu überdenken. Mit Blick auf die Ebene nutzt die Lodge das Prinzip der Wärmemasse, um die Temperaturen das ganze Jahr über auszugleichen und auf eine mechanische Klimatisierung zu verzichten. Sie verfügt auch über eine Reihe von nachhaltigen Innovationen wie zementfreie Bauweise, recycelte Granitböden und solare Wassererwärmung. Insgesamt dient die Lodge als Zeugnis dafür, dass traditionelle Bautechniken wie Stampflehm auf moderne Weise neu interpretiert werden können, um innovative und umweltfreundliche Architektur zu schaffen.

Le lodge est un bâtiment architecturalement unique qui s'intègre parfaitement dans son environnement naturel grâce à l'utilisation de techniques de construction vernaculaires traditionnelles et d'un design minimaliste moderne. Son objectif est de repenser le concept de maison/lodge en utilisant des pratiques et technologies respectueuses de l'environnement. Surplombant les plaines, le lodge utilise le principe de la masse thermique pour aplanir les températures tout au long de l'année, évitant ainsi la climatisation mécanique. Il dispose également d'une gamme d'innovations durables, telles que la construction sans ciment, les sols en granit recyclé et le chauffage solaire de l'eau. Dans l'ensemble, le lodge sert de témoignage à l'idée que les techniques de construction traditionnelles, telles que la terre battue, peuvent être réinterprétées de manière moderne pour créer une architecture innovante et respectueuse de l'environnement.

El lodge es un edificio arquitectónicamente único que se integra perfectamente en su entorno natural mediante el uso de técnicas de construcción vernáculas tradicionales y un diseño minimalista moderno. Su objetivo es repensar el concepto de casa/lodge utilizando prácticas y tecnologías ecológicas que sean respetuosas con el medio ambiente. Con vistas a las llanuras, el lodge utiliza el principio de la masa térmica para aplanar las temperaturas durante todo el año evitando el aire acondicionado mecánico. También cuenta con una gama de innovaciones sostenibles, como la construcción sin cemento, pisos de granito reciclado y calefacción solar de agua. En general, el lodge sirve como un testimonio de que las técnicas de construcción tradicionales como la tierra compactada pueden reinterpretarse de manera moderna para crear una arquitectura innovadora y respetuosa con el medio ambiente.

Elevations

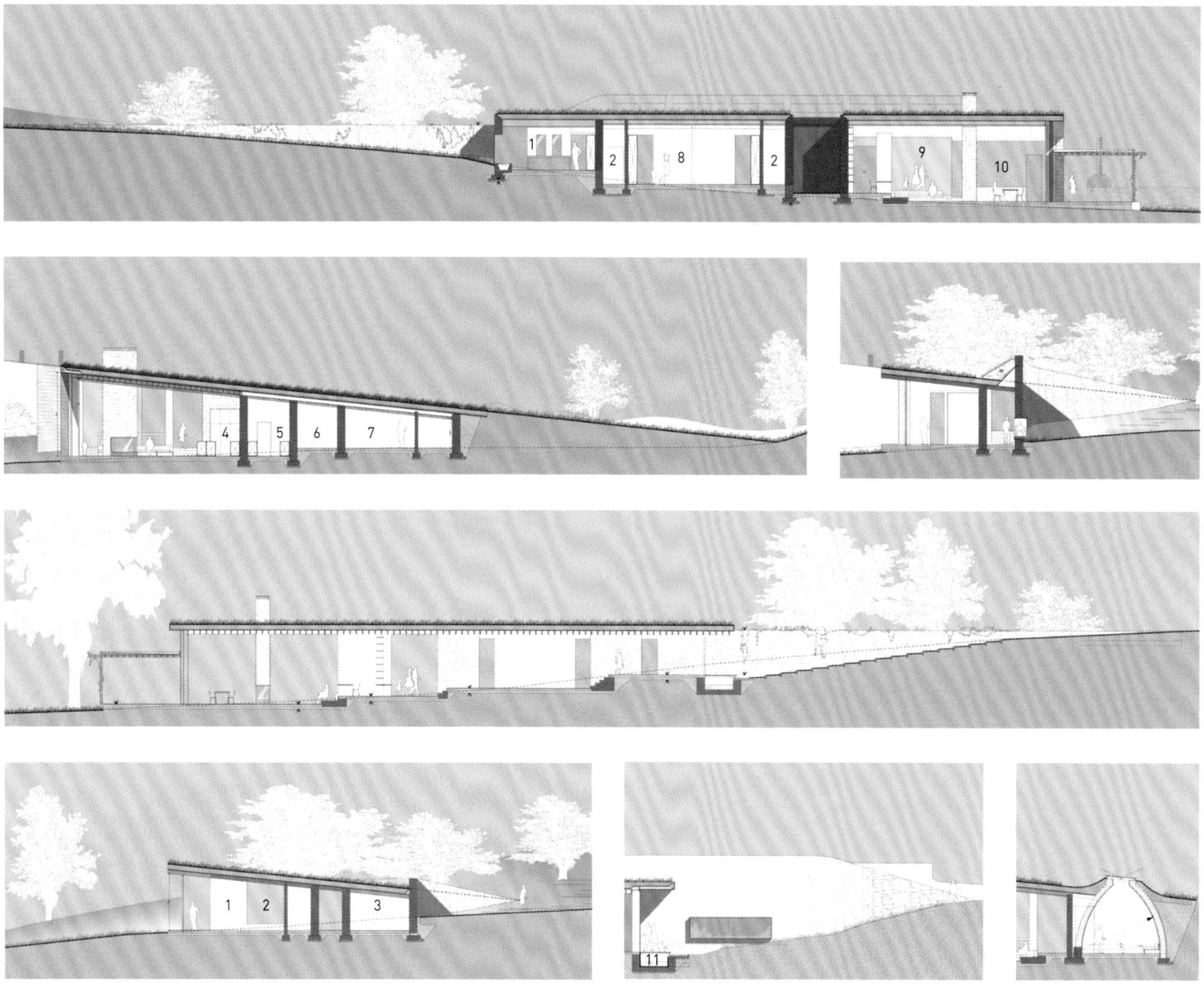

Sections

1. Bedroom
2. Bathroom
3. Store
4. Kitchen
5. Back of kitchen
6. Technical room
7. Maid room
8. Kids bedroom
9. Living room
10. Dining room
11. Drain water tank

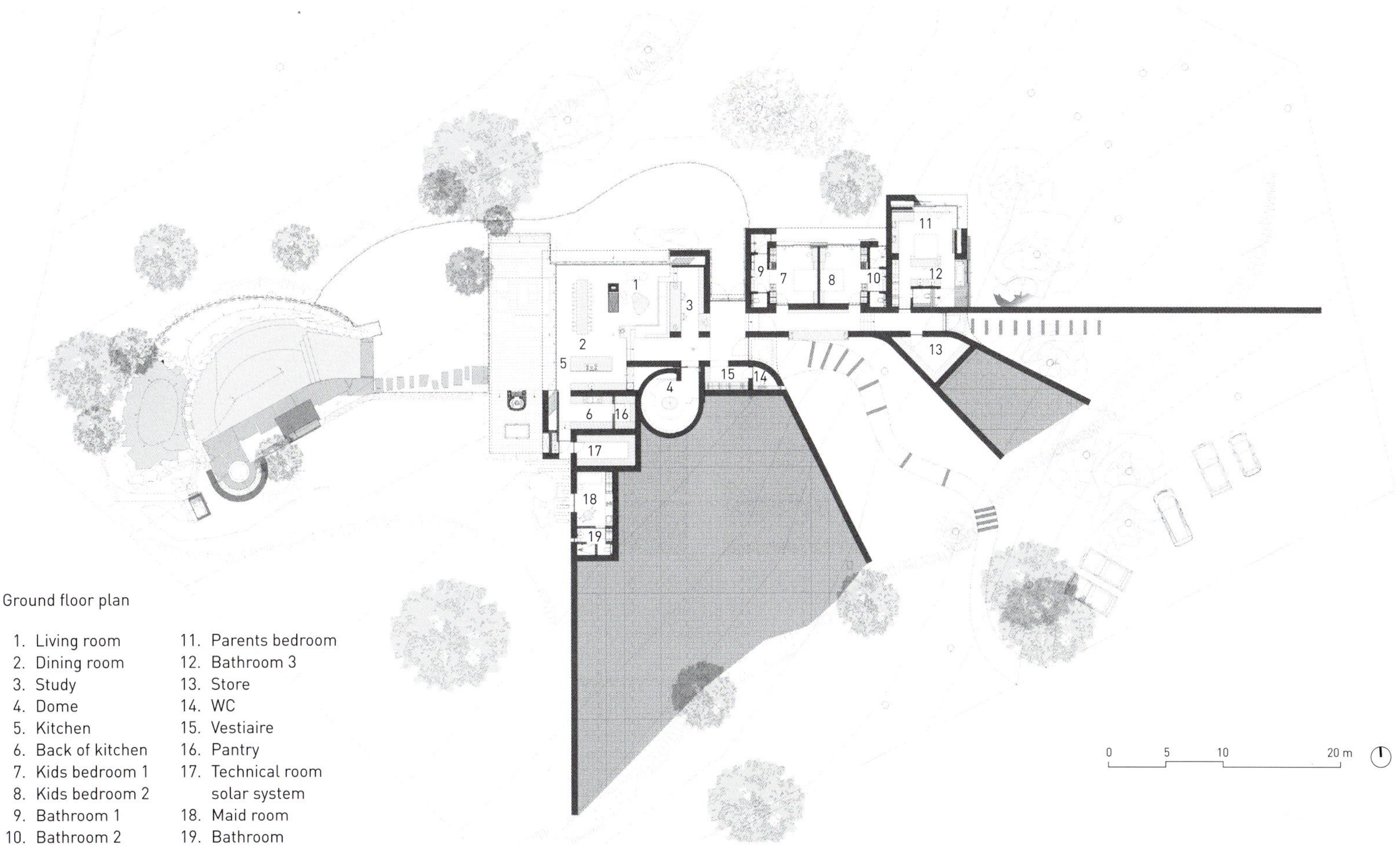

Ground floor plan

1. Living room
2. Dining room
3. Study
4. Dome
5. Kitchen
6. Back of kitchen
7. Kids bedroom 1
8. Kids bedroom 2
9. Bathroom 1
10. Bathroom 2
11. Parents bedroom
12. Bathroom 3
13. Store
14. WC
15. Vestiaire
16. Pantry
17. Technical room solar system
18. Maid room
19. Bathroom

Justin Dingwall
The Vulnerables (diptych)
Photographic Giclee print on
100 % cotton fine art paper
2013

Chris Soal
The Fourth Circle
(The demise of Frank Lucas)
Toothpicks, Burnt & Unburnt,
Polyurethane Sealant,
Industrial Fabric, Board
2019

Founded in 2004, IDIN Architects stands for Integrating Design Into Nature. This design philosophy emphasizes the merge of different senses of surroundings, or "natures," with architectural aesthetics. This is achieved through the analysis and prioritization of each project's needs and requirements. "IDIN" also refers to the beautiful scent after rainfall, symbolizing Thailand's tropical climate, which is a key element in all IDIN designs. Mr. Jeravej Hongsakul, the founding principal, received his Bachelor of Architecture degree from King Mongkut's Institute of Technology Ladkrabang in 1998 and has been practicing architecture since 2000. The company has won numerous national and international awards, including the ASA Architectural Design Awards in Thailand, The Plan Award, Architizer A+ Awards, American Architecture Prize, Arcasia Awards for Architecture, 2A Asia Architecture Award, Design for Asia Awards, German Design Award, GOOD Design Award, and The Prix Versailles South Asia and the Pacific 2020 by UNESCO and the International Union of Architects.

Gegründet im Jahr 2004 steht IDIN Architects für „Integrating Design Into Nature." Diese Design-Philosophie betont die Verschmelzung verschiedener Sinne der Umgebung oder „Natur" mit der architektonischen Ästhetik. Dies wird durch die Analyse und Priorisierung der Bedürfnisse und Anforderungen jedes Projekts erreicht. „IDIN" bezieht sich auch auf den schönen Duft nach Regen, der Thailands tropisches Klima symbolisiert, das ein Schlüsselelement in allen IDIN-Designs ist. Herr Jeravej Hongsakul, der Gründungspartner, erhielt seinen Bachelor of Architecture-Abschluss vom King Mongkut's Institute of Technology Ladkrabang im Jahr 1998 und praktiziert seit 2000 Architektur. Das Unternehmen hat zahlreiche nationale und internationale Auszeichnungen gewonnen, darunter die ASA Architectural Design Awards in Thailand, The Plan Award, Architizer A+ Awards, American Architecture Prize, Arcasia Awards for Architecture, 2A Asia Architecture Award, Design for Asia Awards, German Design Award, GOOD Design Award und The Prix Versailles South Asia and the Pacific 2020 von UNESCO und der International Union of Architects.

TARA VILLA

Fondé en 2004, IDIN Architects signifie « Integrating Design Into Nature ». Cette philosophie de conception met l'accent sur la fusion des différents sens de l'environnement, ou « natures », avec l'esthétique architecturale. Cela est réalisé grâce à l'analyse et à la priorisation des besoins et exigences de chaque projet. « IDIN » fait également référence à la belle odeur après la pluie, symbolisant le climat tropical de la Thaïlande, qui est un élément clé dans tous les designs d'IDIN. M. Jeravej Hongsakul, le fondateur principal, a obtenu son diplôme de baccalauréat en architecture du King Mongkut's Institute of Technology Ladkrabang en 1998 et pratique l'architecture depuis 2000. L'entreprise a remporté de nombreux prix nationaux et internationaux, notamment les ASA Architectural Design Awards en Thaïlande, The Plan Award, les Architizer A+ Awards, l'American Architecture Prize, les Arcasia Awards for Architecture, les 2A Asia Architecture Award, les Design for Asia Awards, le German Design Award, le GOOD Design Award et le Prix Versailles South Asia and the Pacific 2020 par l'UNESCO et l'Union internationale des architectes.

Fundada en 2004, IDIN Architects significa «Integrando el Diseño en la Naturaleza». Esta filosofía de diseño enfatiza la fusión de diferentes sentidos del entorno, o «naturalezas», con la estética arquitectónica. Esto se logra a través del análisis y la priorización de las necesidades y requisitos de cada proyecto. «IDIN» también se refiere al hermoso aroma después de la lluvia, simbolizando el clima tropical de Tailandia, que es un elemento clave en todos los diseños de IDIN. El Sr. Jeravej Hongsakul, el fundador y principal arquitecto, obtuvo su Licenciatura en Arquitectura en el King Mongkut's Institute of Technology Ladkrabang en 1998 y ha estado practicando la arquitectura desde 2000. La compañía ha ganado numerosos premios nacionales e internacionales, incluidos los Premios de Diseño Arquitectónico ASA en Tailandia, el Premio Plan, los Premios Architizer A+, el Premio de Arquitectura Americana, los Premios Arcasia para la Arquitectura, el Premio de Arquitectura 2A Asia, los Premios de Diseño para Asia, el Premio de Diseño Alemán, el Premio de Diseño BUENO y el Prix Versailles del Sur de Asia y el Pacífico 2020 por la UNESCO y la Unión Internacional de Arquitectos.

TARA VILLA

KANCHANABURI, THAILAND // PHOTOS © DOF SKY|GROUND

Start - Complete Date: 2014 - 20 december 2019
Usable Area: 5,900 m^2
Owner: Pramote Kamolpantip
Architect: IDIN Architects (Jeravej Hongsakul, Eakgaluk Sirijariyawat, Wichan Kongnok)
Interior Designer: IDIN Architects Jureerat Korvanichakul, Thuwanont Ruangkanoksilp, Poowadon Siriburanakij
Structural Engineer: Pakanut Siriprasopsothorn
System Engineer: Eakachai Hamhomvong, Panot Kuakoolwong
Contractor: ArtCon

Situated at the Kwai riverbank, a small resort serves as a lively event host for weddings, conferences, and seminars, with facilities that offer a full cycle of event programs. Despite the relatively small number of guestrooms compared to the site area, the designer paid more attention to the space between each functional area and utilized the idea that the journey is more important than the destination as source material. The concept is represented through the human senses, including touch, sight, hearing, and smell, which become an integral part of the users' experience while they walk through the resort. The building layout connects the main circulation with every common area, while the villas serve as the heart of the resort. The rammed earth walls, made from leftover soil, become the main feature of the resort and interact with the users and their senses. Tara Villa exemplifies the efficient utilization of material and space, with a design process that emphasizes the spaces between functions rather than the physical structure.

Ein kleines Resort am Ufer des Kwai dient als lebhafter Veranstaltungsort für Hochzeiten, Konferenzen und Seminare und bietet Einrichtungen für den gesamten Zyklus von Veranstaltungsprogrammen. Trotz der vergleichsweise geringen Anzahl von Gästezimmern im Vergleich zur Grundstücksfläche widmete der Designer mehr Aufmerksamkeit dem Raum zwischen den einzelnen Funktionsbereichen und nutzte die Idee, dass die Reise wichtiger ist als das Ziel als Quellmaterial. Das Konzept wird durch die menschlichen Sinne repräsentiert, einschließlich Berührung, Sehen, Hören und Geruch, die zu einem integralen Bestandteil der Erfahrung der Benutzer werden, während sie durch das Resort spazieren gehen. Die Gebäudestruktur verbindet die Hauptzirkulation mit jedem gemeinsamen Bereich, während die Villen das Herzstück des Resorts bilden. Die stampflehmwände, die aus übrig gebliebenem Boden hergestellt werden, sind das Hauptmerkmal des Resorts und interagieren mit den Benutzern und ihren Sinnen. Tara Villa verkörpert die effiziente Nutzung von Material und Raum, mit einem Designprozess, der den Raum zwischen den Funktionen betont und nicht die physische Struktur.

Situé sur la rive de la rivière Kwai, un petit complexe hôtelier sert de lieu d'accueil animé pour des mariages, des conférences et des séminaires, avec des installations proposant un cycle complet de programmes d'événements. Malgré le nombre relativement restreint de chambres par rapport à la superficie du site, le designer a accordé plus d'attention à l'espace entre chaque zone fonctionnelle et a utilisé l'idée que le voyage est plus important que la destination comme matériau source. Le concept est représenté à travers les sens humains, y compris le toucher, la vue, l'ouïe et l'odorat, qui deviennent une partie intégrante de l'expérience des utilisateurs lorsqu'ils se promènent dans le complexe. La disposition des bâtiments relie la circulation principale à chaque zone commune, tandis que les villas servent de cœur du complexe hôtelier. Les murs en pisé, fabriqués à partir de sol restant, deviennent la caractéristique principale du complexe et interagissent avec les utilisateurs et leurs sens. Tara Villa incarne l'utilisation efficace du matériau et de l'espace, avec un processus de conception qui met l'accent sur les espaces entre les fonctions plutôt que sur la structure physique.

Ubicado en la orilla del río Kwai, un pequeño resort sirve como un animado anfitrión de eventos para bodas, conferencias y seminarios, con instalaciones que ofrecen un ciclo completo de programas de eventos. A pesar del número relativamente pequeño de habitaciones de huéspedes en comparación con el área del sitio, el diseñador prestó más atención al espacio entre cada área funcional y utilizó la idea de que el viaje es más importante que el destino como material fuente. El concepto se representa a través de los sentidos humanos, incluyendo el tacto, la vista, el oído y el olfato, que se convierten en una parte integral de la experiencia de los usuarios mientras caminan por el resort. El diseño del edificio conecta la circulación principal con cada área común, mientras que las villas sirven como el corazón del resort. Las paredes de tierra apisonada, hechas con tierra sobrante, se convierten en la característica principal del resort e interactúan con los usuarios y sus sentidos. Tara Villa ejemplifica la utilización eficiente de material y espacio, con un proceso de diseño que enfatiza los espacios entre las funciones en lugar de la estructura física.

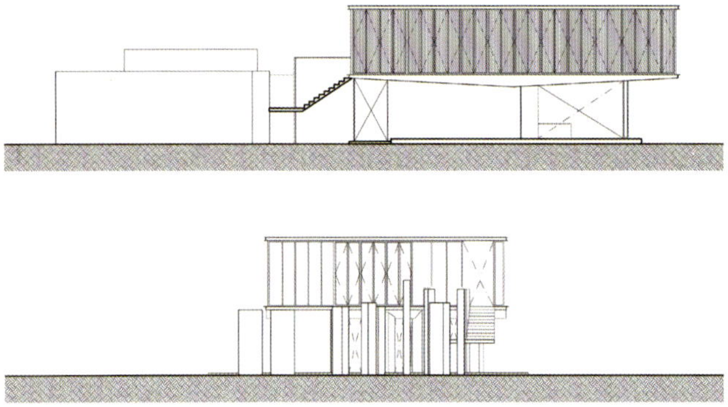

Main lobby. Elevations

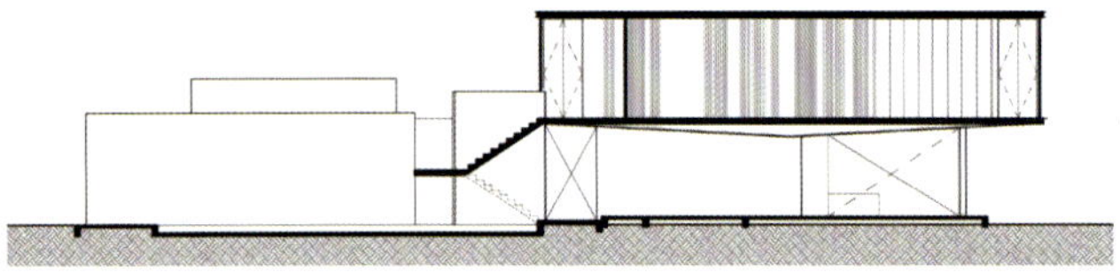

Main lobby. Section

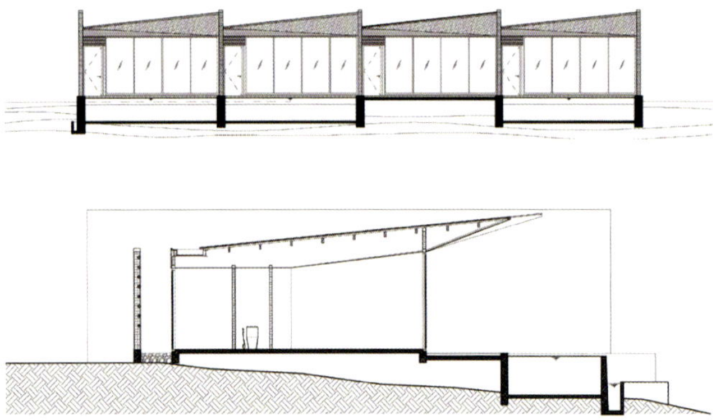

Pool villa. Sections

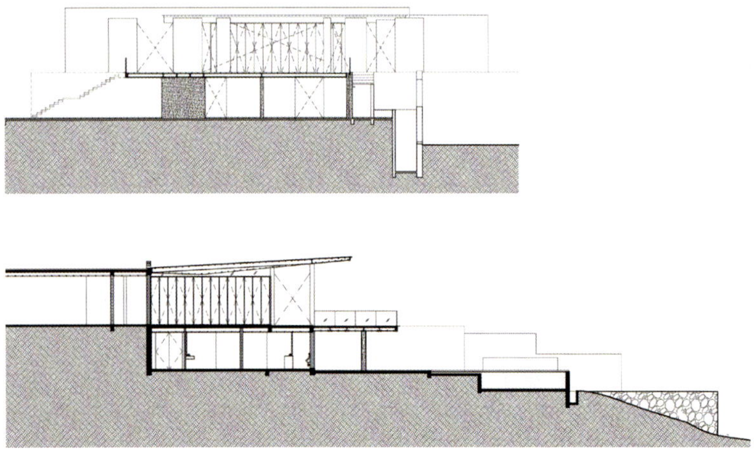

Restaurant. Sections

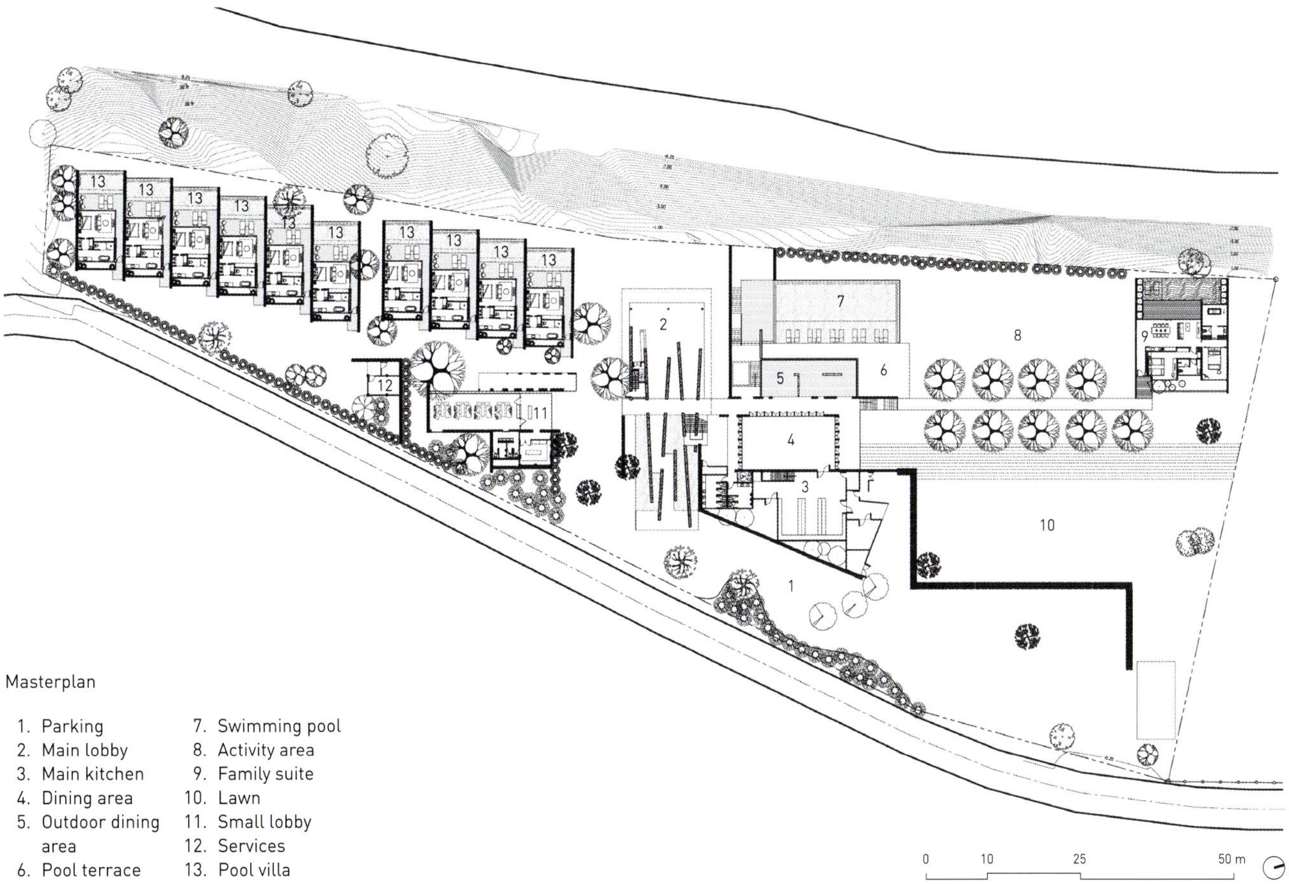

Masterplan

1. Parking
2. Main lobby
3. Main kitchen
4. Dining area
5. Outdoor dining area
6. Pool terrace
7. Swimming pool
8. Activity area
9. Family suite
10. Lawn
11. Small lobby
12. Services
13. Pool villa

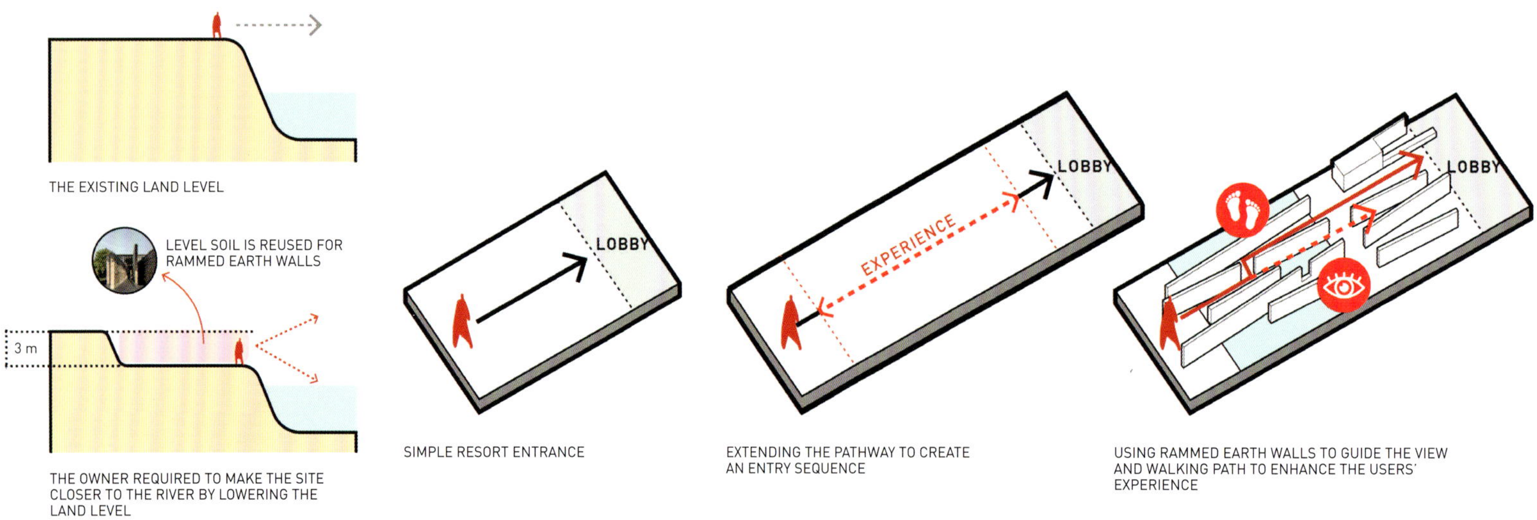

Conceptual diagrams

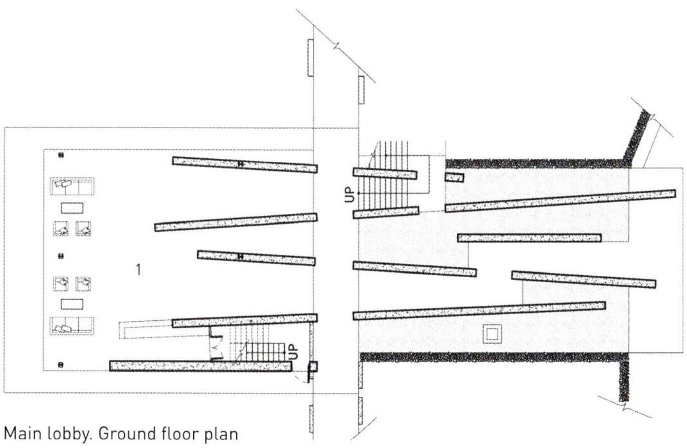

Main lobby. Ground floor plan

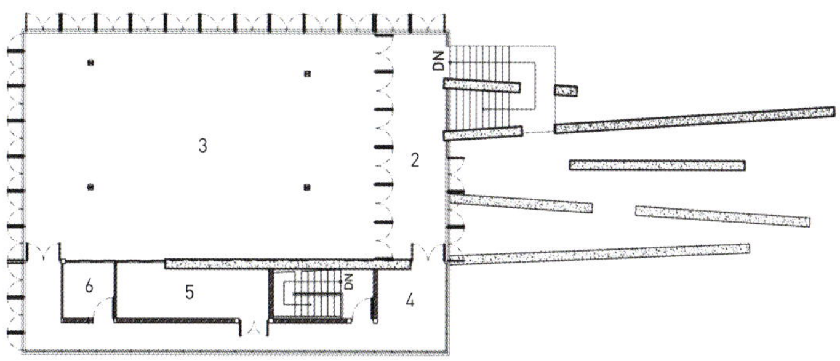

Main lobby. First floor plan

1. Lobby lounge
2. Foyer
3. Function room
4. Service corridor
5. Storage room
6. M&E room

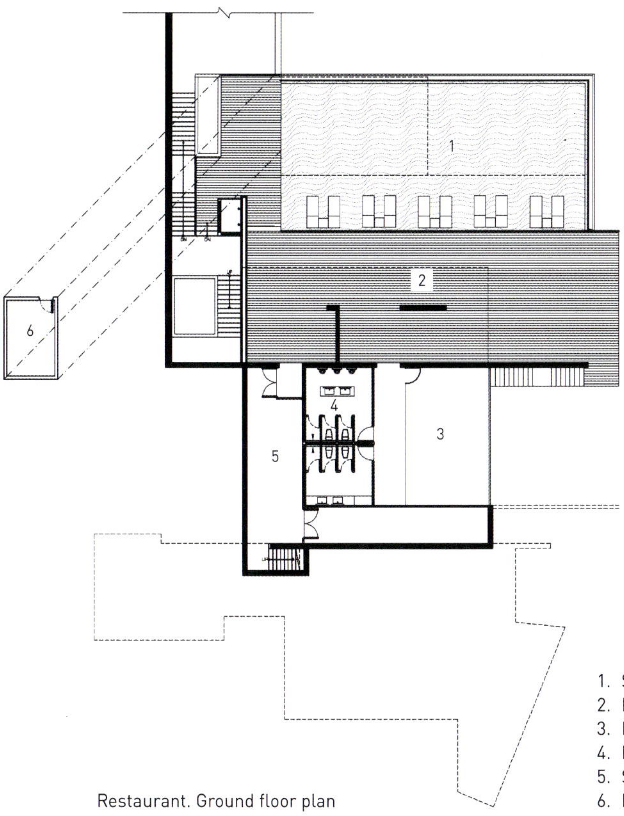

Restaurant. Ground floor plan

1. Swimming pool
2. Pool terrace
3. Fitness room
4. Restroom
5. Storage
6. Pool pump

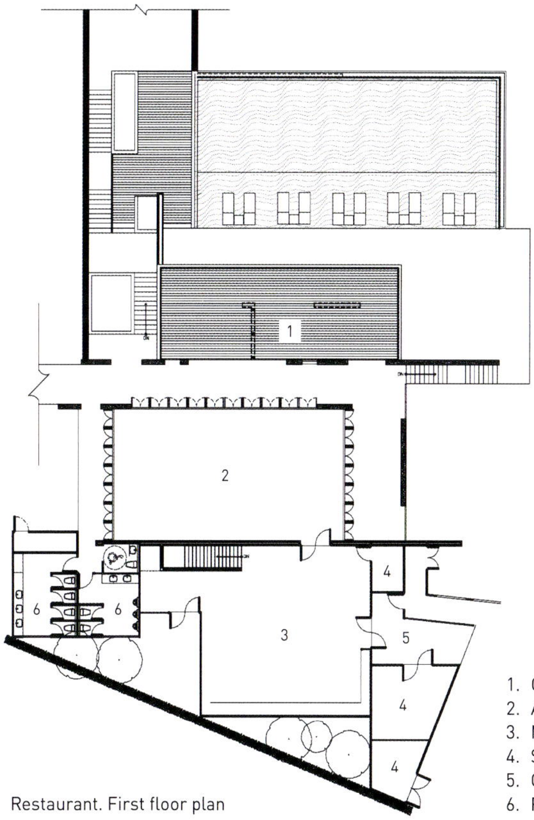

Restaurant. First floor plan

1. Outdoor dining
2. All-day dining
3. Main kitchen
4. Storage
5. Corridor
6. Restroom

Restaurant. First floor plan

1. Bedroom
2. Living area
3. Bathroom
4. Terrace
5. Swimming pool

David Kaunitz and Ka Wai Yeung are a husband-wife team focused on facilitating high quality architecture in some of the most disadvantaged communities and First Nations People in Australian and Asia – Pacific. Underlying this is a deep commitment to participatory design and local construction. They have created a new paradigm that shows high quality, international award-winning architecture is not a matter of cost, but of commitment. All their work has been delivered within the same funding and time constraints as other similar projects. This has reset the benchmark of what is possible and highlighted that high quality, change making architecture does not need to be a luxury item. They have a significant body of award-winning architecture and have worked in more than 40 Aboriginal & Torres Strait Island communities, 200 Pacific Island communities as well as in Asia. This includes prestigious awards such as the Union of International Architects 2021 Vassilis Sgoutas Prize, the worlds highest architectural honor for working with underprivileged communities.

David Kaunitz und Ka Wai Yeung sind ein Ehepaar, das qualitativ hochwertige Architektur in einigen der am benachteiligsten Communities und First Nations in Australien und im asiatisch-pazifischen Raum ermöglichen möchte. Hinter dieser Philosophie steckt eine tiefe Verpflichtung zu partizipativem Design und lokalem Bauen. Sie haben ein neues Paradigma geschaffen, das zeigt, dass hochwertige, international ausgezeichnete Architektur keine Frage der Kosten, sondern des Engagements ist. Die gesamte Arbeit wurde unter den gleichen finanziellen und zeitlichen Einschränkungen wie bei anderen ähnlichen Projekten durchgeführt. Dies hat die Messlatte des Möglichen neu gesetzt und deutlich gemacht, dass hochwertige, nachhaltige Architektur kein Luxusgut sein muss. Sie verfügen einen bedeutenden Bestand an preisgekrönter Architektur und haben in über 40 Aborigine- und Torres Strait Islander-Gemeinschaften, 200 pazifischen Inselgemeinschaften und Asien gearbeitet. Dazu gehören prestigeträchtige auszeichnungen wie der Vassilis Sgoutas Award 2021 der International Union of Architects, die weltweit höchste architektonische auszeichnung für die arbeit mit benachteiligten gemeinden.

PUNTUKURNU ABORIGINAL HEALTH SERVICE HEALTHCARE HUB

David Kaunitz et Ka Wai Yeung sont un couple qui se consacre à la promotion d'une architecture de qualité dans certaines des communautés les plus défavorisées et des premières nations d'Australie et d'Asie-Pacifique. Cette approche repose sur un engagement profond en faveur de la conception participative et de la construction locale. Ils ont créé un nouveau paradigme qui démontre que l'architecture de haute qualité, récompensée au niveau international, n'est pas une question de coût, mais d'engagement. Tous leurs travaux ont été réalisés dans le cadre des mêmes contraintes de financement et de temps que d'autres projets similaires. Cela a permis de redéfinir les critères de référence de ce qui est possible et de montrer clairement qu'une architecture de haute qualité et porteuse de changement ne doit pas être un article de luxe. Ils possèdent un important corpus d'architecture primée et ont travaillé dans plus de 40 communautés aborigènes et insulaires du détroit de Torres, 200 communautés insulaires du Pacifique et en Asie. Cela inclut des récompenses prestigieuses telles que le prix Vassilis Sgoutas 2021 de l'Union internationale des architectes, la plus haute distinction architecturale au monde pour le travail avec les communautés défavorisées.

David Kaunitz y Ka Wai Yeung son un equipo de marido y mujer centrado en facilitar una arquitectura de alta calidad en algunas de las comunidades más desfavorecidas y en las Primeras Naciones de Australia y Asia-Pacífico. Subyace un profundo compromiso con el diseño participativo y la construcción local. Han creado un nuevo paradigma que demuestra que la arquitectura de alta calidad, premiada internacionalmente, no es una cuestión de coste, sino de compromiso. Todo su trabajo se ha realizado con las mismas limitaciones de financiación y tiempo que otros proyectos similares. Esto ha reajustado el punto de referencia de lo que es posible y ha puesto de manifiesto que la arquitectura de alta calidad y capaz de generar cambios no tiene por qué ser un artículo de lujo. Tienen un importante corpus de arquitectura premiada y han trabajado en más de 40 comunidades aborígenes y de las islas Torres Strait, en 200 comunidades de las islas del Pacífico y en Asia. Esto incluye prestigiosos premios como el Premio Vassilis Sgoutas de la Unión Internacional de Arquitectos 2021, el más alto honor arquitectónico del mundo por trabajar con comunidades desfavorecidas.

PUNTUKURNU ABORIGINAL HEALTH SERVICE HEALTHCARE HUB

NEWMAN, WESTERN AUSTRALIA // PHOTOS © ROBERT FRITH, ACORN PHOTO

AWARDS:

- Australian Institute of Architects Winner 2020
- NSW Reconciliation Prize (www.architecture.com.au/prizes/nsw/reconciliation-prize Body of Work including this project)

Funding Body: Australian Department of Health
Client: Robby Chibawe, CEO Puntukurnu Aboriginal Medical Service
Client Project Managers: Guy Docker, Matthew Sanderson
Source Art Centre: Martumili Art Centre
Artists: Nyarrie Morgan, Nancy Nyanjilpayi Chapman, Dadda Sampson, Bugai Whyoulter, Sue Bung
The Collaborative Artwork by: Bugai Whyoulter, Dulcie Gibbs- Jugarda, Jakayu Biljabu, Janice Nixon- Yuwali, Kumpaya Girgirba, Mabel Mitutu- Wakarta, Marjorie Yates, Morika Biljabu, Ngamaru Bidu, Nola Taylor- Ngalangka, Nora Nungabar, Nora Wompi, Rosie Williams- Muntararr, and Thelma Judson
Landscape Architect: Conrad Gargett
Structural / Building Services: Prompt Engineering
Civil Engineer: Stellen Consulting
QS: QS Services
Traffic Engineering: KCTT
Compliance: Tecon
Contractors: Devlyn
Solar Panels: Geniux
Visualisation Sketches: Andrew Andersons
Construction Period: May 2019 – June 2020
Total Built Area: 900 m^2

This primary health care centre is the first to be built in Newman, a town with a majority Aboriginal population, and houses the Puntukurnu Aboriginal Medical Service (PAMS), which serves communities up to 800 km away. Its design was based on a participatory process that involved the community and resulted in an architecture more in keeping with the place, the Aboriginal people and their culture. Similarly, the design of the project integrates creations by 19 artists representing the 5 communities served by the centre. The building is organised around a courtyard that separates the health clinic from the administrative area, generating at the same time a space for interaction and visual connection. The rammed earth walls from the site that form the building envelope create an intuitive connection with the local population. In addition, the public park created around the centre, with 2000 low-maintenance local endemic plants, is instrumental in contextualising and connecting the building to the region. The project includes a photovoltaic installation that provides 100 % of the building's electricity under optimal sunlight conditions and more than 90 % of all energy.

Dieses Zentrum für die medizinische Grundversorgung ist das erste, das in Newman, einer Stadt mit einer mehrheitlichen Aborigine-Bevölkerung, gebaut wurde, und beherbergt den Puntukurnu Aboriginal Medical Service (PAMS), der Gemeinden im Umkreis bis zu 800 km versorgt. Der Entwurf basierte auf einem partizipativen Prozess, der die Gemeinde einbezog und zu einer Architektur führte, die dem Ort, den Aborigines und ihrer Kultur besser entspricht. Auf die gleiche Weise integriert das Design des Projekts Werke von 19 Künstlern, die die 5 Gemeinden repräsentieren, die das Zentrum bedient. Das Gebäude ist um einen Innenhof herum organisiert, der die Gesundheitsklinik vom Verwaltungsbereich trennt und gleichzeitig einen Raum für Interaktion und visuelle Verbindung schafft. Die Stampflehmwände aus dem Gelände, die die Gebäudehülle bilden, schaffen eine intuitive Verbindung mit der lokalen Bevölkerung. Ebenso ist der um das Zentrum herum angelegte öffentliche Park mit 2.000 pflegeleichten, einheimischen Pflanzen von grundlegender Bedeutung für die Kontextualisierung und Verbindung des Gebäudes mit der Region. Das Projekt beinhaltet eine Photovoltaikanlage, die bei optimaler Sonneneinstrahlung 100 % des Stroms des Gebäudes und mehr als 90 % der gesamten Energie liefert.

Ce centre de soins de santé primaires est le premier à être construit à Newman, une ville dont la population est majoritairement indigène, et abrite le service médical aborigène Puntukurnu (PAMS), qui dessert des communautés situées jusqu'à 800 km de là. Sa conception s'est appuyée sur un processus participatif qui a impliqué la communauté et a abouti à une architecture plus en accord avec le site, le peuple autochtone et sa culture. De même, la conception du projet intègre les créations de 19 artistes représentant les 5 communautés desservies par le centre. Le bâtiment est organisé autour d'une cour qui sépare la clinique de la zone administrative, générant en même temps un espace d'interaction et de connexion visuelle. Les murs en pisé du site qui forment l'enveloppe du bâtiment créent une connexion intuitive avec la population locale. De même, le parc public créé autour du centre, avec 2 000 plantes endémiques locales nécessitant peu d'entretien, est fondamental pour contextualiser et relier le bâtiment à la région. Le projet comprend une installation photovoltaïque qui fournit 100 % de l'électricité du bâtiment dans des conditions d'ensoleillement optimales et plus de 90 % de l'énergie totale.

Este centro de salud de atención primaria es el primero construido en Newman, una localidad con población de mayoría aborigen, y alberga el Servicio Médico Aborigen Puntukurnu (PAMS), que atiende a comunidades situadas a una distancia de hasta 800 km. Su diseño se basó en un proceso participativo que implicó a la comunidad y que dio como resultado una arquitectura más acorde con el lugar, la población aborigen y su cultura. De la misma forma, el diseño del proyecto integra creaciones de 19 artistas que representan a las 5 comunidades a las que da servicio el centro. El edificio se organiza alrededor de un patio que separa la clínica de salud de la zona administrativa, generando al mismo tiempo un espacio de interacción y conexión visual. Los muros de tierra apisonada procedente del lugar que conforman la envolvente del edificio crean una conexión intuitiva con la población local. Así mismo, el parque público creado alrededor del centro, con 2000 plantas endémicas locales de bajo mantenimiento, resulta fundamental para contextualizar y conectar el edificio con la región. El proyecto incluye una instalación fotovoltaica que proporciona el 100 % de la electricidad del edificio en condicionas óptimas de asoleo y más del 90 % de toda la energía.

View from east

View from south east

Sketches by Andrew Andersons AO

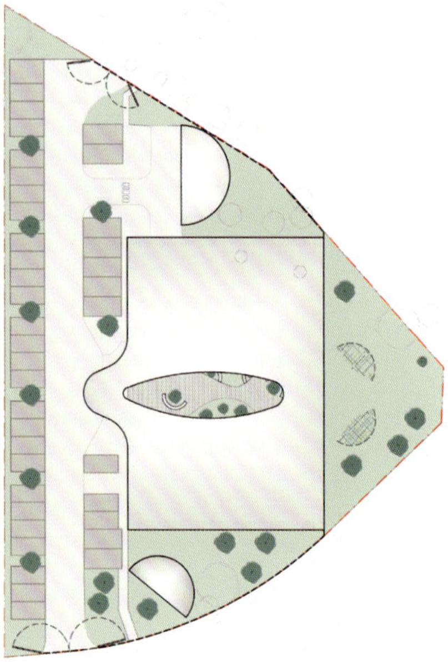

Site plan

Ground floor plan

The founder of Le House Company is the architect Le Hung Trong (19/11/1983): before of the establishment in 2015, it took him nine years to get an Archviz Artist position in an architecture firm after he graduated from Van Lang University - Architecture Design.
In the next seven years, he was alone on his way to look for the differences in how to conceive design. When he realized completely that he had enough confidence to start up his own company based on his passions and experiences, he decided to do it with those wishing to contribute to Le House's creativity to present the image of Viet Nam's architecture internationally.
"To be creative is forgetting the definitions of what is already in your mind."

Der Gründer von Le House Company ist der Architekt Le Hung Trong (19.11.1983): Vor der Gründung im Jahr 2015 brauchte er neun Jahre, um nach seinem Abschluss an der Van Lang University - Architecture Design - eine Stelle als Archviz Artist in einem Architekturbüro zu bekommen.
In den folgenden sieben Jahren war er allein auf der Suche nach den Unterschieden in der Art und Weise, wie man Design konzipiert. Als er feststellte, dass er genug Selbstvertrauen hatte, um sein eigenes Unternehmen auf der Grundlage seiner Leidenschaften und Erfahrungen zu gründen, beschloss er, dies gemeinsam mit denjenigen zu tun, die zur Kreativität von Le House beitragen wollten, um das Image der vietnamesischen Architektur international zu präsentieren.
„Kreativ zu sein bedeutet, die Definitionen dessen, was man bereits im Kopf hat, zu vergessen."

KASAYA – VẠT CÀ SA - VEGETARIAN RESTAURANT

Le fondateur de Le House Company est l'architecte Le Hung Trong (19/11/1983) : avant la création en 2015, il lui a fallu neuf ans pour obtenir un poste d'Archviz Artist dans un cabinet d'architecture après avoir été diplômé de l'Université Van Lang - Architecture Design.
Au cours des sept années suivantes, il était seul sur son chemin à la recherche des différences dans la façon de concevoir le design. Lorsqu'il s'est rendu compte qu'il avait suffisamment confiance en lui pour créer sa propre entreprise en s'appuyant sur ses passions et ses expériences, il a décidé de le faire avec ceux qui souhaitent contribuer à la créativité de Le House pour présenter l'image de l'architecture vietnamienne à l'échelle internationale.
« Être créatif, c'est oublier les définitions de ce qui est déjà dans votre esprit ».

El fundador de Le House Company es el arquitecto Le Hung Trong (19/11/1983): antes de la creación en 2015, tardó nueve años en conseguir un puesto de Artista en un estudio de arquitectura tras graduarse en la Universidad Van Lang - Diseño de Arquitectura.
En los siguientes siete años, se concentró en su camino para buscar las diferencias en la forma de concebir el diseño. Cuando se dio cuenta por completo de que tenía la suficiente confianza para crear su propia empresa basada en sus pasiones y experiencias, decidió hacerlo con quienes deseaban contribuir a la creatividad de Le House para presentar la imagen de la arquitectura de Vietnam a nivel internacional.
«Ser creativo es olvidar las definiciones de lo que ya está en tu mente».

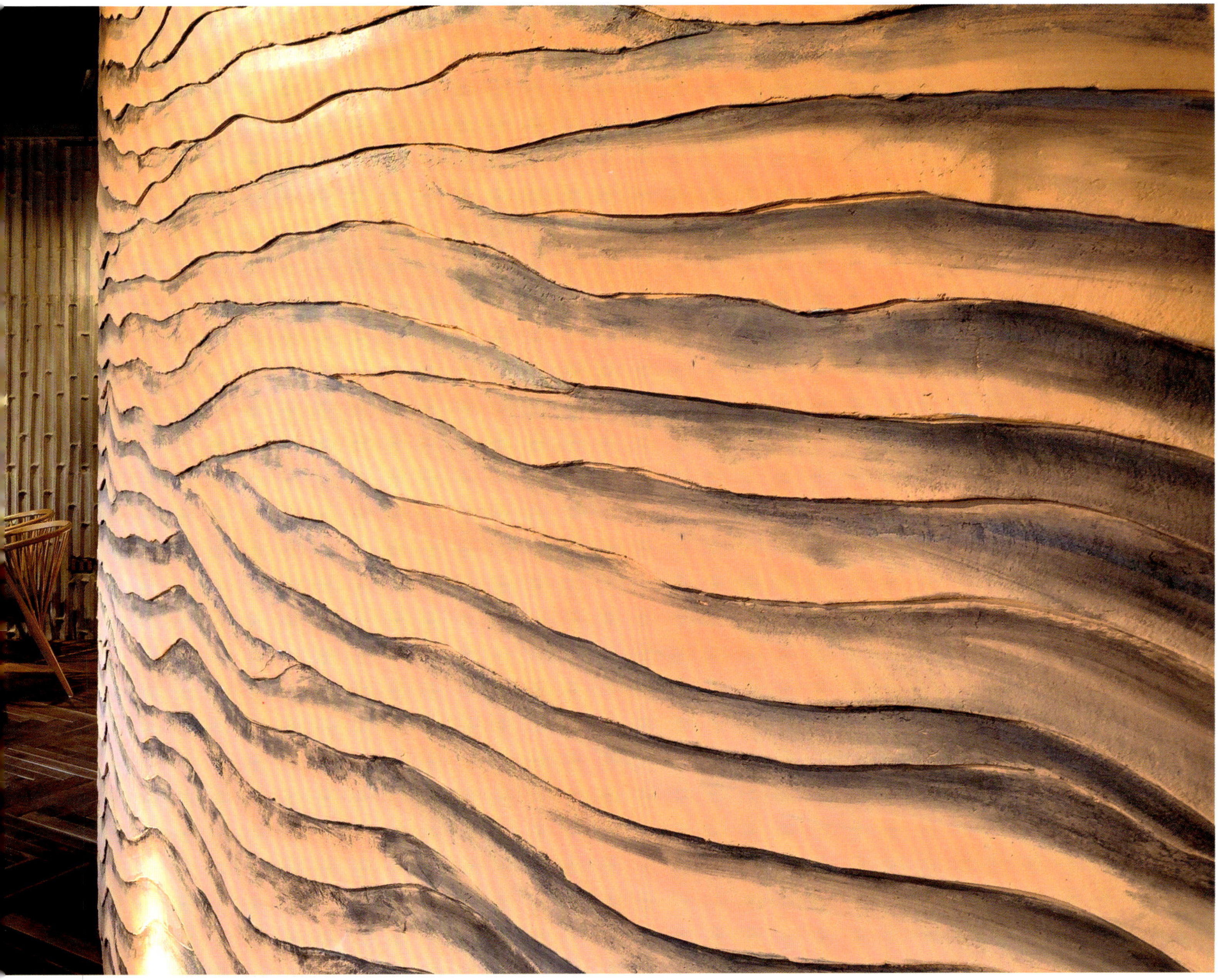

KASAYA – VẠT CÀ SA - VEGETARIAN RESTAURANT

BA ĐÌNH DISTRICT, HÀ NỘI, VIETNAM // PHOTOS © TRIỆU CHIẾN

Project Year: 2022
Area: 1600 m²
Designed by: Le House
Construction by: Thuyan construction
M&A Art Tiles: black-handmade ceramic
Jumbo Coating: Custom metal art, design & Fabrication
Thuyan: interior construction
Betong Decor: concrete grinding platform
Golden Bee-Art Solution: Facade

The name Kasaya is inspired by the Sanskrit translation of "kashāya," which means the robes of the Buddhist monk. It is called "Vạt Cà Sa" in Vietnamese. In Buddhism, the kashāya is a symbol of holy life, virtue and the golden light of divine enlightenment.The building is located on a plot of 10 x 20 m (around 200 m²), with 8 floors, of which 4 are restaurants, 1 is a kitchen block and the remaining 3 are the living space of a 3-generation family. In addition, the location is in Hà Nội Old Quarter and with a rather prominent height, which inspired the architect to exploit the design form in the Vertical azimuth. The unique architecture stands out with the shape of the kashāya running smoothly from the top and stretching down to the entrance of the restaurant. The steel structure stairs beautified by the CNC cutting model with bamboo leaf pattern is also one of the typical symbols of the restaurant. The success of the project lies in the continuous opening and closing of space, creating a connection throughout the height of the restaurant.

Der Name Kasaya ist inspiriert von der Sanskrit-Übersetzung des Wortes „kashāya," was die Roben des buddhistischen Mönchs bedeutet. Auf Vietnamesisch wird es „Vạt Cà Sa" genannt. Im Buddhismus symbolisiert die kashāya das heilige Leben, die Tugend und das goldene Licht der allmächtigen Erleuchtung.Das Gebäude befindet sich auf einem Grundstück von 10 x 20 m (etwa 200 m²) und verfügt über 8 Etagen, von denen 4 Restaurants sind, 1 Etage eine Küchenblock ist und die verbleibenden 3 Etagen der Wohnraum einer 3-Generationen-Familie sind. Außerdem befindet sich das Gebäude im alten Stadtviertel von Hà Nội und hat eine ziemlich markante Höhe, was den Architekten dazu inspirierte, die Form im vertikalen Azimut zu gestalten. Die einzigartige Architektur zeichnet sich durch die Form der kashāya aus, die vom oberen Teil bis zum Eingang des Restaurants glatt verläuft. Die Stahlstruktur der Treppe ist mit dem CNC-Schnittmuster mit Bambusblattmuster verziert, ebenfalls ein typisches Symbol des Restaurants. Der Erfolg des Projekts liegt in der kontinuierlichen Öffnung und Schließung des Raums, wodurch eine Verbindung entlang der Höhe des Restaurants entsteht.

Le nom Kasaya est inspiré de la traduction sanskrit de « kashāya », qui signifie les robes du moine bouddhiste. Il est appelé « Vạt Cà Sa » en vietnamien. Dans le bouddhisme, la kashāya est un symbole de la vie sainte, de la vertu et de la lumière dorée de l'illumination divine.Le bâtiment est situé sur un terrain de 10 x 20 m (environ 200 m²), avec 8 étages, dont 4 sont des restaurants, 1 est un bloc de cuisine et les 3 autres sont l'espace habitable d'une famille de 3 générations.De plus, l'emplacement se trouve dans le vieux quartier de Hà Nội et avec une hauteur assez remarquable, ce qui a inspiré l'architecte à exploiter la forme de conception dans l'azimut vertical. L'architecture unique se distingue par la forme de la kashāya qui s'étend du haut jusqu'à l'entrée du restaurant. La structure en acier des escaliers est embelli par le modèle de découpe CNC avec un motif de feuille de bambou, également l'un des symboles typiques du restaurant. Le succès du projet réside dans l'ouverture et la fermeture continues de l'espace, créant une connexion le long de la hauteur du restaurant.

El nombre Kasaya está inspirado en la traducción sánscrita de «kashāya», que significa las túnicas del monje budista. Se llama «Vạt Cà Sa» en vietnamita. En el budismo, la kashāya es un símbolo de la vida sagrada, la virtud y la luz dorada de la iluminación divina.El edificio está ubicado en un terreno de 10 x 20 m (alrededor de 200 m²), con ocho pisos, de los cuales cuatro son restaurantes, uno es un bloque de cocina y los tres restantes son el espacio habitable de una familia de tres generaciones. Además, la ubicación se encuentra en el barrio antiguo de Hà Nội y con una altura bastante destacada, lo que inspiró al arquitecto a explotar la forma de diseño en la azimut vertical. La arquitectura única se destaca con la forma de la kashāya que se extiende desde la parte superior hasta la entrada del restaurante. La estructura de acero de las escaleras está embellecida por el modelo de corte CNC con patrón de hojas de bambú, también uno de los símbolos típicos del restaurante. El éxito del proyecto radica en la apertura y cierre continuo del espacio, creando una conexión a lo largo de la altura del restaurante.

Front elevation

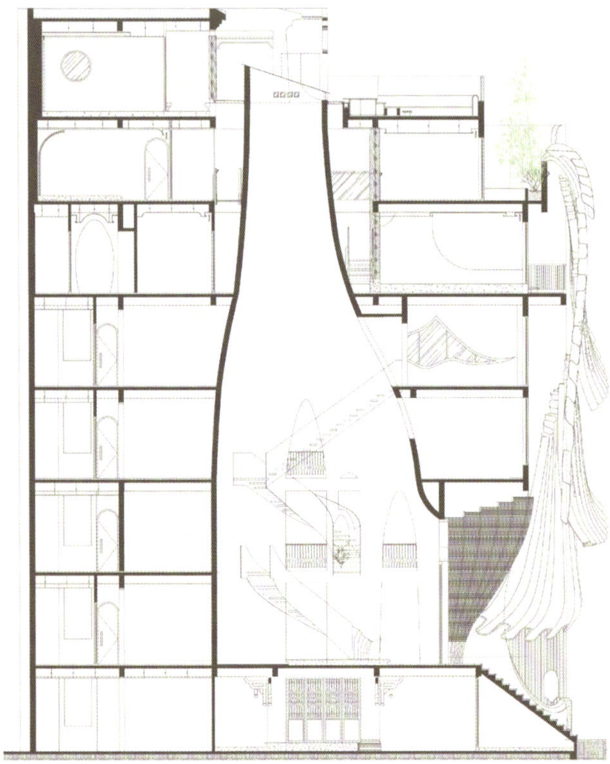

Section

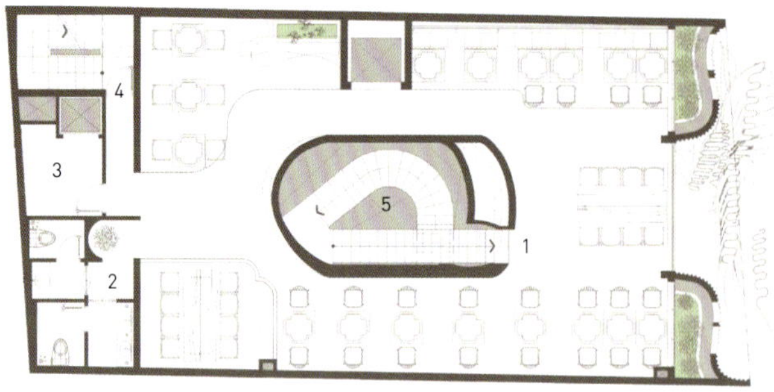

Fourth floor plan

1. Entrance
2. Stairway
3. Food preparation room
4. Stairwell
5. Void

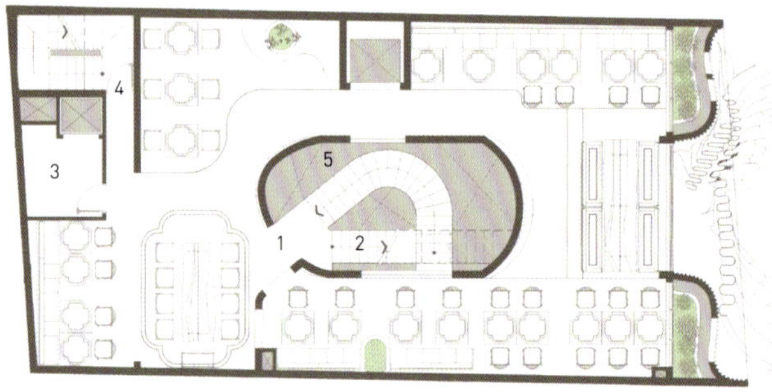

Third floor plan

1. Entrance
2. Stairway
3. Food preparation room
4. Stairwell
5. Void

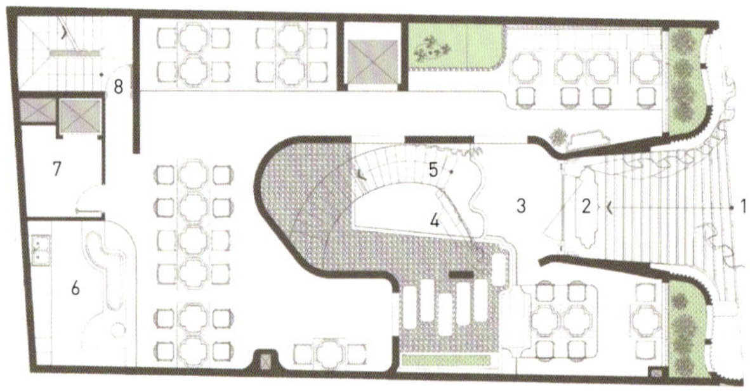

First floor plan

1. Main entrance
2. Lobby
3. Reception hall
4. Reception
5. Stairway
6. Bar counter
7. Food preparation room
8. Stairwell

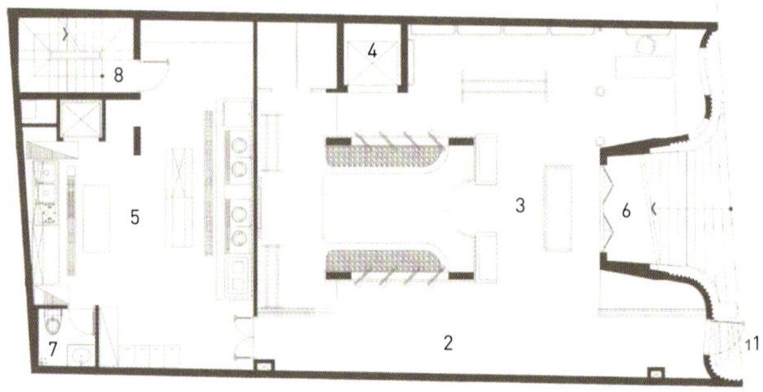

First floor plan

1. Side entrance
2. Lobby
3. Ao Dai Showroom
4. Elevator
5. Toilet
6. Stairwell
7. Toilet
8. Stairwell

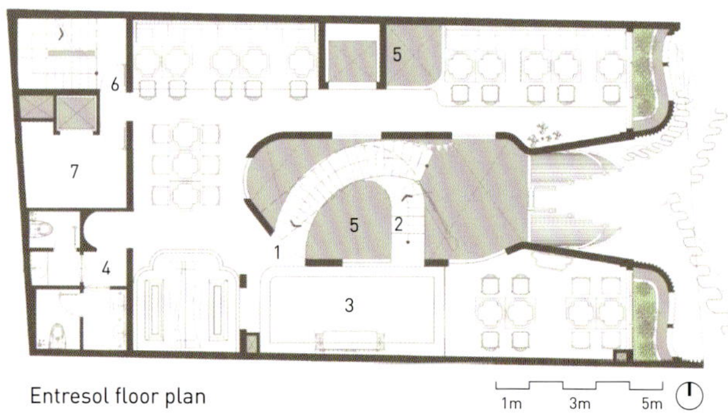

Entresol floor plan

1. Entrance
2. Stairway
3. Lobby
4. Toilet
5. Void
6. Stairwell
7. Food preparation room

Lucila Aguilar Arquitectos is a firm in Mexico City that strives to serve the earth through an architecture that inspires others to dream and innovate in order to achieve a socially conscious and sustainable future. It aims to find balance and harmony with nature through designs that use natural, renewable materials such as bamboo and earth; exploring diverse techniques that respond to the unique conditions of each site. Through the work of Lucila Aguilar Arquitectos, the firm has positioned itself as a pioneer in bamboo and earth architecture in Mexico.
Lucila Aguilar founded the firm in 2016. Since then, the firm has received recognitions such as an Honorific Mention in the XIV Mexican Architecture Biennial, Finalist in the Architizer A+Awards, runner up un the World Design Awards, amongst others. It was invited to participate in the 2019 Biennial in La Habana, the 2021 Biennial in Venice and to exhibit at INK Talk India 2020 as well as to design the Pavilion for Design Week Mexico 2020. The firm aspires to demonstrate with creativity and innovation that natural materials and artisan processes can be the solution for a promising future.

Lucila Aguilar Arquitectos ist ein Architekturbüro in Mexiko-Stadt, das durch Architektur, die andere dazu inspiriert, zu träumen und zu innovieren, um eine nachhaltige Zukunft mit sozialem Bewusstsein zu erreichen, der Erde dienen möchte. Durch die Suche nach Balance und Harmonie mit der Natur durch Entwürfe, die natürliche erneuerbare Materialien wie Bambus und Erde verwenden; Erkundung verschiedener Techniken, die auf die einzigartigen Bedingungen des Standorts reagieren. Durch die Arbeit von Lucila Aguilar Arquitectos hat sich das Büro als Vorreiter in der Bambusarchitektur in Mexiko positioniert.
Lucila Aguilar gründete das Büro 2016. Seitdem hat sie Auszeichnungen wie die Ehrenvolle Erwähnung bei der XIV. Biennale für Mexikanische Architektur, Finalistin bei den Architizer A + Awards, Runner-up bei den World Design Awards und andere erhalten. Sie wurde eingeladen, an der Biennale von Havanna 2019, der Biennale von Venedig 2021 teilzunehmen und auf der INK Talk India 2020 auszustellen. Sie wurde eingeladen, den Pavillon für die Design Week Mexico 2020 zu entwerfen. Sie strebt danach, mit Kreativität und Innovation zu zeigen, dass natürliche Materialien und handwerkliche Prozesse die Lösung für eine vielversprechende Zukunft sein können.

PERICOS

Lucila Aguilar Arquitectos est un cabinet d'architectes situé dans la ville de Mexico qui cherche à servir la terre à travers une architecture qui inspire les autres à rêver et à innover pour atteindre un avenir durable avec une conscience sociale. En trouvant l'équilibre et l'harmonie avec la nature, à travers des conceptions qui utilisent des matériaux naturels renouvelables tels que le bambou et la terre; en explorant différentes techniques qui répondent aux conditions uniques du site. Grâce au travail de Lucila Aguilar Arquitectos, le cabinet s'est positionné comme un pionnier de l'architecture en bambou au Mexique.
Lucila Aguilar a fondé le cabinet en 2016. Depuis lors, elle a reçu des prix tels que la Mention honorable à la XIVe Biennale d'architecture mexicaine, Finaliste aux Architizer A + Awards, finaliste aux World Design Awards, entre autres. Elle a été invitée à participer à la Biennale de La Havane 2019, à la Biennale de Venise 2021 et à exposer lors de la conférence INK Talk India 2020. Elle a été invitée à concevoir le pavillon pour la Design Week Mexico 2020. Elle aspire à démontrer avec créativité et innovation que les matériaux naturels et les processus artisanaux peuvent être la solution pour un avenir prometteur.

Lucila Aguilar Arquitectos es un despacho en la Ciudad de México que busca servir a la tierra a través de una arquitectura que inspire a otros a soñar e innovar para alcanzar un futuro sostenible con conciencia social. Encontrando el balance y armonía con la naturaleza, a través de diseños con materiales naturales, renovables como bambú y tierra; explorando diversas técnicas que responden a las condiciones únicas del sitio. A través del trabajo de Lucila Aguilar Arquitectos, el despacho se ha posicionado como pionero en la arquitectura de bambú en México.
Lucila Aguilar fundó el despacho en 2016. Desde entonces ha recibido premios como Mención Honorífica en la XIV Bienal de Arquitectura Mexicana, Finalista en Architizner A+Awards, runner up en los World Design Awards, entre otros. Fue invitada a participar en La Bienal de La Habana 2019, Bienal de Venecia 2021 y exponer en INK Talk India 2020. Fue invitada a diseñar el Pabellón para Design Week México 2020. Aspira a demostrar con creatividad e innovación que los materiales naturales y procesos artesanales pueden ser la solución para un futuro prometedor.

PERICOS

SOCONUSCO REGION, CHIAPAS, MEXICO
PHOTOS © LUCILA AGUILAR DE LA LAMA

Design Architect: Lucila Aguilar Arquitectos
Client: Finca Don Jorge
Principal in Charge: Lucila Aguilar de la Lama
Bamboo Constructor: Bolivar Epigmenio Garcia Miranda
Design Team: Miguel Ruiz Velasco, Habid Athie, Ricardo Romero, Paola Calzada
Administration: María Teresa Moreno
Civil Engineer: Cesar Moreno Peralta

Finca Don Jorge is located in Mexico Soconusco Region, in a state-of-the-art 110-hectare cacao plantation. The compound consists of 5 modules with 9 rooms that have an impressive view of the Tacaná volcano. The sustainable design leverages from the natural materials available in the area: Earth, Bamboo, Wood, River Rocks. The adobe walls are separated from the humidity of the ground with river stone. The bamboo and esterilla roof is covered in vegetation to maintain a cool temperature. The "Pericos" rooms blend in with nature and visitors can enjoy the birds songs and the peaceful atmosphere while reconnecting with Mexican cacao and it´s ancient legacy.

Die Finca Don Jorge befindet sich in der Soconusco-Region in Mexiko auf einer modernen 110 Hektar großen Kakao-Plantage. Die Anlage besteht aus 5 Modulen mit 9 Zimmern, die einen beeindruckenden Blick auf den Tacaná-Vulkan bieten. Das nachhaltige Design nutzt die natürlichen Materialien, die in der Gegend zur Verfügung stehen: Erde, Bambus, Holz, Flussgestein. Die Adobe-Wände sind von der Feuchtigkeit des Bodens durch Flusssteine getrennt. Das Bambus- und Esterilladach ist mit Vegetation bedeckt, um eine kühle Temperatur zu erhalten. Die „Pericos"-Zimmer fügen sich in die Natur ein und die Besucher können die Vogelgesänge und die friedliche Atmosphäre genießen und sich dabei mit dem mexikanischen Kakao und seinem alten Erbe verbinden.

Finca Don Jorge est située dans la région Soconusco du Mexique, dans une plantation de cacao de pointe de 110 hectares. Le complexe se compose de 5 modules avec 9 chambres qui offrent une vue impressionnante sur le volcan Tacaná. Le design durable tire parti des matériaux naturels disponibles dans la région : terre, bambou, bois, roches de rivière. Les murs en adobe sont séparés de l'humidité du sol par des pierres de rivière. Le toit en bambou et esterilla est recouvert de végétation pour maintenir une température fraîche. Les chambres « Pericos » se fondent dans la nature et les visiteurs peuvent profiter du chant des oiseaux et de l'atmosphère paisible tout en renouant avec le cacao mexicain et son ancienne tradition.

La Finca Don Jorge, ubicada en la Región Soconusco de México, dentro de una plantación de 110 hectáreas de cacao de última generación. El conjunto consta de 5 módulos con 9 habitaciones que disfrutan de una impresionante vista del volcán Tacaná. El diseño sustentable se pensó para aprovechar los materiales naturales disponibles en la zona: Tierra, Bambú, Madera, Piedra Bola. Sus muros de adobe están protegidos de la humedad del suelo con piedras de río. El techo de bambú y esterilla está cubierto de vegetación para mantener una temperatura fresca. Las habitaciones «Pericos» se integran con el paisaje para que los huéspedes pueden disfrutar del canto de las aves y la paz del ambiente, mientras se reconectan con el cacao mexicano, un gran legado ancestral.

Constructive section

1. VEGETATION
   Insulating cover
2. BAMBOO
   Structure
3. WOOD
   Floors
4. STONE
   Superbasement
5. EARTH
   Adobes

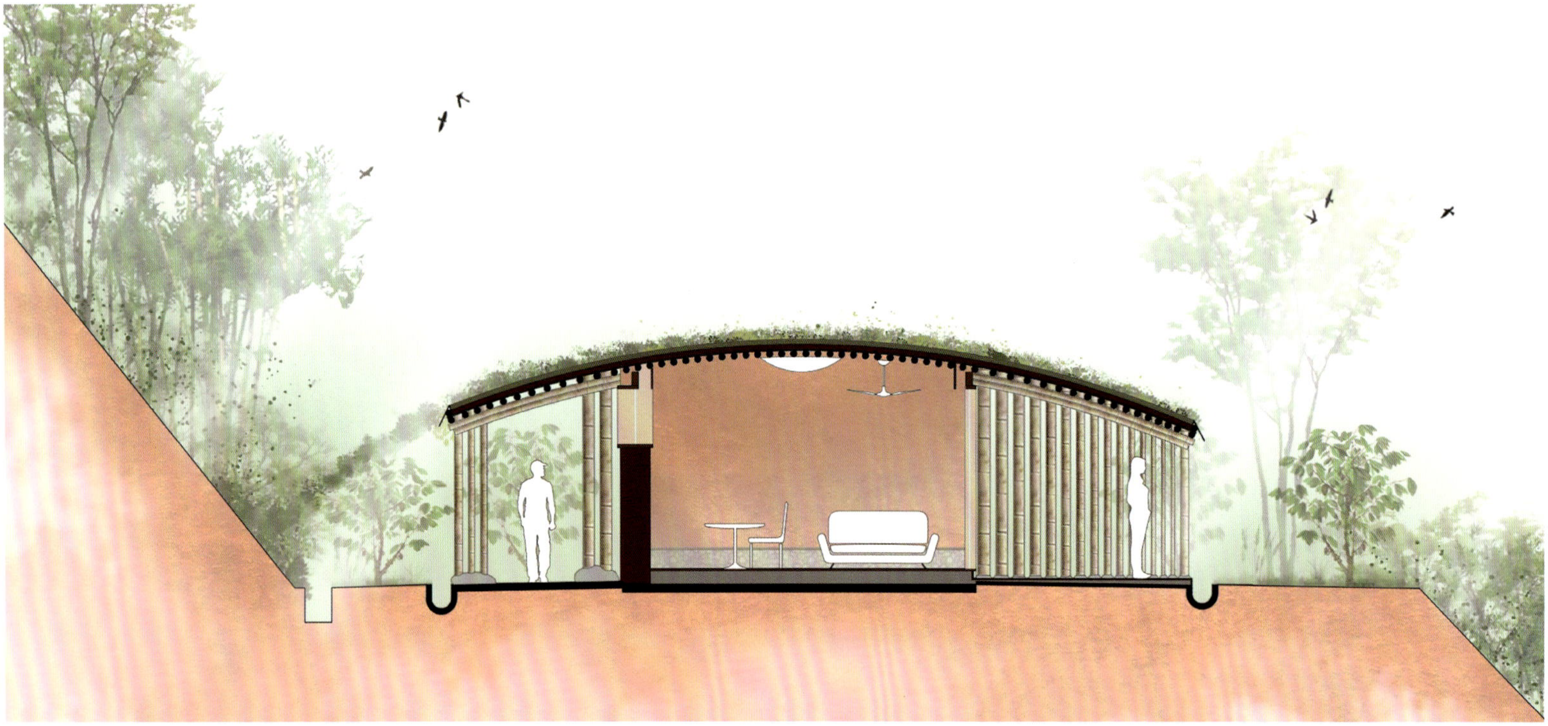

General section

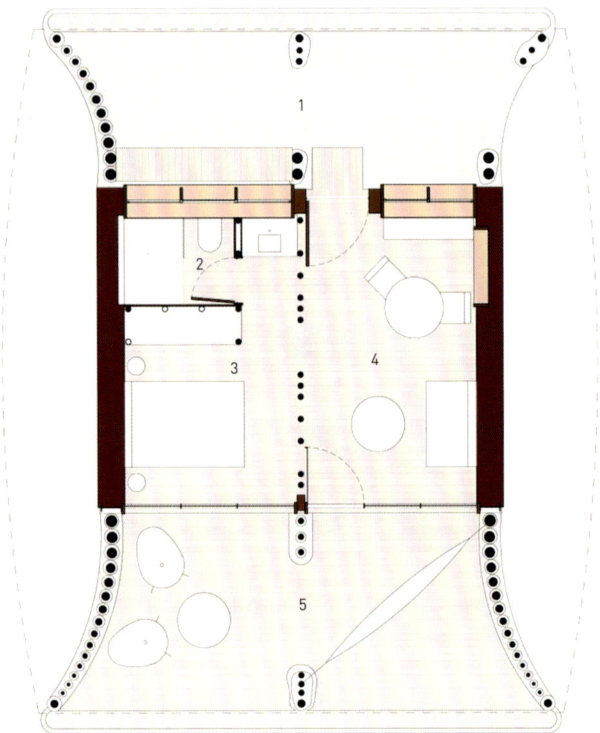

Suite module. Floor plan

1. Access portico
2. Bathroom
3. Bedroom
4. Living area
5. Terrace

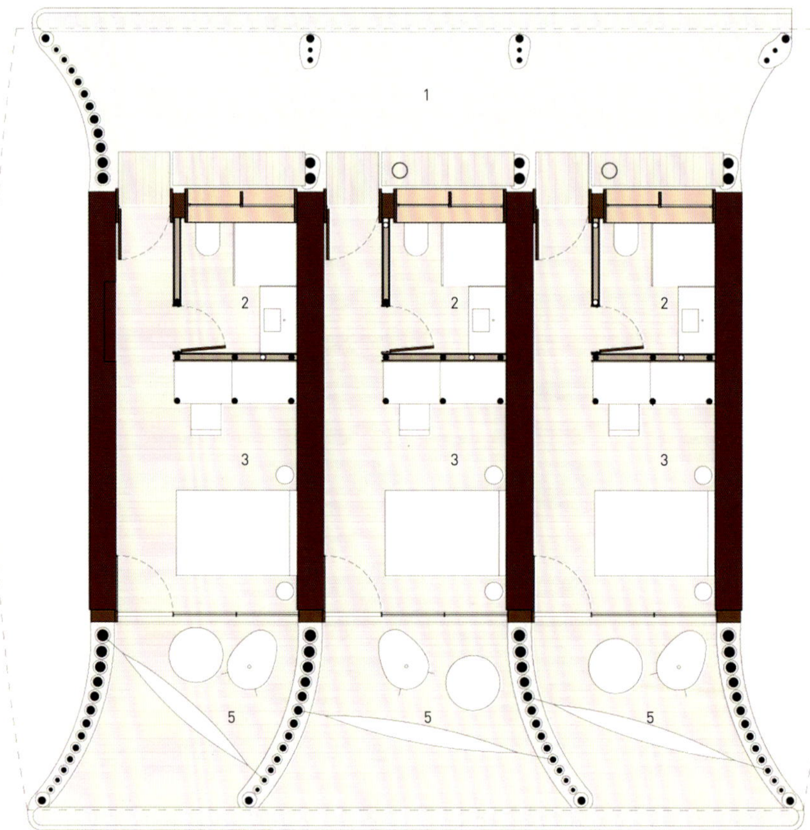

Bedroom module. Floor plan

Site plan

Nativa Arquitectura is a Mexican company that was established in 2018 in Baja California. The firm specializes in implementing local and native resources in a contemporary way to create architectural practices in the region. Its design process is based on understanding the site to create proposals that integrate into their surroundings, taking advantage of the innate characteristics of each location. From material proposals to site regeneration, Nativa Arquitectura seeks harmony in each project. The firm's philosophy is concrete and focuses on preserving and reinterpreting traditional construction techniques. Each project is an opportunity to explore the cultural richness and natural beauty of the Baja California region. In summary, Nativa Arquitectura is dedicated to creating innovative designs that respect the cultural heritage and biodiversity of the region.

Nativa Arquitectura ist ein mexikanisches Unternehmen, das 2018 in Baja California gegründet wurde. Die Firma spezialisiert sich auf die Umsetzung von lokalen und einheimischen Ressourcen auf zeitgenössische Weise, um architektonische Praktiken in der Region zu schaffen. Ihr Designprozess basiert auf dem Verständnis des Standorts, um Vorschläge zu erstellen, die sich in die Umgebung integrieren und die natürlichen Eigenschaften jedes Ortes nutzen. Von der Materialvorschlägen bis zur Regeneration des Ortes strebt Nativa Arquitectura Harmonie in jedem Projekt an. Die Philosophie des Unternehmens ist konkret und konzentriert sich darauf, traditionelle Bautechniken zu bewahren und neu zu interpretieren. Jedes Projekt ist eine Gelegenheit, den kulturellen Reichtum und die natürliche Schönheit der Region Baja California zu erkunden. Zusammenfassend widmet sich Nativa Arquitectura der Schaffung innovativer Designs, die das kulturelle Erbe und die Biodiversität der Region respektieren.

CASA MARTHA

Nativa Arquitectura est une entreprise mexicaine fondée en 2018 en Baja California. La société se spécialise dans la mise en œuvre de ressources locales et autochtones de manière contemporaine pour créer des pratiques architecturales dans la région. Son processus de conception repose sur la compréhension du site pour créer des propositions qui s'intègrent à leur environnement, en exploitant les caractéristiques innées de chaque lieu. De la proposition de matérialité à la régénération du site, Nativa Arquitectura recherche l'harmonie dans chaque projet. La philosophie de l'entreprise est concrète et vise à préserver et réinterpréter les techniques de construction traditionnelles. Chaque projet est une opportunité d'explorer la richesse culturelle et la beauté naturelle de la région de Baja California. En résumé, Nativa Arquitectura se consacre à la création de designs innovants qui respectent le patrimoine culturel et la biodiversité de la région.

Nativa Arquitectura es una empresa mexicana que se estableció en 2018 en Baja California. La firma se especializa en la implementación de recursos locales y nativos de forma contemporánea para crear prácticas arquitectónicas en la región. Su proceso de diseño se basa en la comprensión del sitio para crear propuestas que se integren en su entorno, aprovechando las características innatas de cada lugar. Desde la propuesta de materialidad hasta la regeneración del sitio, Nativa Arquitectura busca la armonía en cada proyecto. La filosofía de la firma es concreta y se enfoca en preservar y reinterpretar las técnicas tradicionales de construcción. Cada proyecto es una oportunidad para explorar la riqueza cultural y la belleza natural de la región de Baja California. En resumen, Nativa Arquitectura se dedica a crear diseños innovadores que respetan el patrimonio cultural y la biodiversidad de la región.

CASA MARTHA

LA MISIÓN, BAJA CALIFORNIA, MEXICO
PHOTOS © OSCAR HERNÁNDEZ RODRÍGUEZ - IG: @OHFA_

Date of Construction: 2021
Architect: Alfredo Navarro Tiznado
Design Team: Kenia Esmeralda García Rosas, Hanna Appel Hernández
Construction: Pedro Luis Curiel Bojórquez, José Francisco Ramírez García

Casa Martha is an archaeologist's home designed with rudimental and artisanal techniques, unifying time, form, and context through composition and materiality of ancestral origin. Located in the urban area of La Misión, between Ensenada and Rosarito, the lot has dimensions of 15 x 40 meters on rugged topography. The program includes a retirement home for a couple, Martha and Bill, and a space for plastic arts. The house is blended with the site, and a dialogue is created between three volumes embedded in the site, oriented towards the Pacific Ocean. The first level has two areas, the visitor area, and the study area. The second level has the entertainment room, dining room, and kitchen, accessed through large windows and porches woven with salt pine. At the back of the social area, a staircase leads to the third level, where the main chamber is located. Casa Martha is modeled with deep sensitivity and respect for its surroundings, using compacted earth as the main construction element.

Casa Martha ist das Zuhause eines Archäologen, das mit rudimentären und handwerklichen Techniken gestaltet wurde und durch Zusammensetzung und Materialität mit ursprünglichem Ursprung Zeit, Form und Kontext vereint. Es befindet sich in der Stadt La Misión zwischen Ensenada und Rosarito, auf einem Grundstück mit Abmessungen von 15 x 40 Metern und einer unebenen Topographie. Das Programm umfasst ein Ruhestandszuhause für ein Paar, Martha und Bill, sowie einen Raum für die Plastikkunst. Das Haus ist in die Umgebung integriert, und es entsteht ein Dialog zwischen drei in das Gelände eingebetteten Volumina, die zum Pazifischen Ozean hin ausgerichtet sind. Die erste Ebene hat zwei Bereiche, den Besucherbereich und den Studienbereich. Die zweite Ebene verfügt über das Unterhaltungszimmer, Esszimmer und die Küche, die durch große Fenster und Veranden aus Salzkiefer betreten werden. Am Ende des sozialen Bereichs führt eine Treppe zur dritten Ebene, auf der sich das Hauptschlafzimmer befindet. Casa Martha wurde mit großer Sensibilität und Respekt für ihre Umgebung modelliert und verwendet verdichtete Erde als Hauptbauelement.

Casa Martha est la maison d'un archéologue conçue avec des techniques rudimentaires et artisanales, unifiant le temps, la forme et le contexte par la composition et la matérialité d'origine ancestrale. Situé dans la zone urbaine de La Misión, entre Ensenada et Rosarito, le terrain a des dimensions de 15 x 40 mètres sur une topographie accidentée. Le programme comprend une maison de retraite pour un couple, Martha et Bill, et un espace pour les arts plastiques. La maison est intégrée au site, et un dialogue est créé entre trois volumes intégrés dans le site, orientés vers l'océan Pacifique. Le premier niveau a deux zones, la zone visiteurs et la zone d'étude. Le deuxième niveau comprend la salle de divertissement, la salle à manger et la cuisine, accessibles par de grandes fenêtres et des porches tissés avec du pin salé. À l'arrière de la zone sociale, un escalier mène au troisième niveau, où se trouve la chambre principale. Casa Martha est modelée avec une profonde sensibilité et un respect pour son environnement, en utilisant la terre compactée comme principal élément de construction.

Casa Martha es la casa de un arqueólogo diseñada con técnicas rudimentarias y artesanales, unificando el tiempo, la forma y el contexto a través de la composición y la materialidad de origen ancestral. Ubicada en el área urbana de La Misión, entre Ensenada y Rosarito, el lote tiene dimensiones de 15 x 40 metros en una topografía accidentada. El programa incluye una casa de retiro para una pareja, Martha y Bill, y un espacio para artes plásticas. La casa se integra con el sitio, y se crea un diálogo entre tres volúmenes incrustados en el sitio, orientados hacia el Océano Pacífico. El primer nivel tiene dos áreas, el área de visitantes y el área de estudio. El segundo nivel tiene la sala de entretenimiento, comedor y cocina, accesibles a través de grandes ventanales y porches tejidos con pino salado. En la parte trasera del área social, una escalera conduce al tercer nivel, donde se encuentra la cámara principal. Casa Martha está modelada con profunda sensibilidad y respeto por su entorno, utilizando tierra compactada como el elemento principal de construcción.

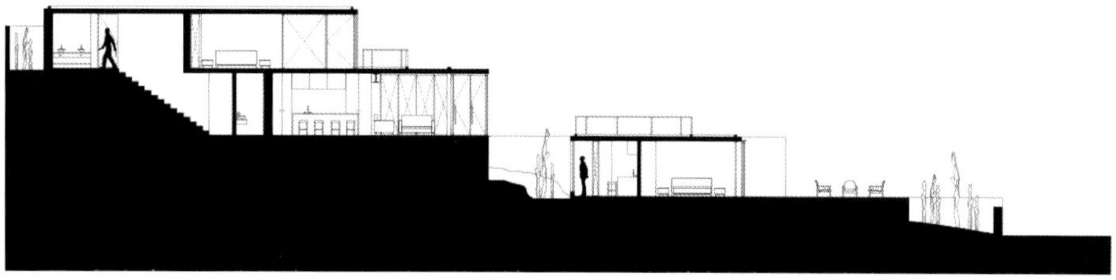

Longitudinal section

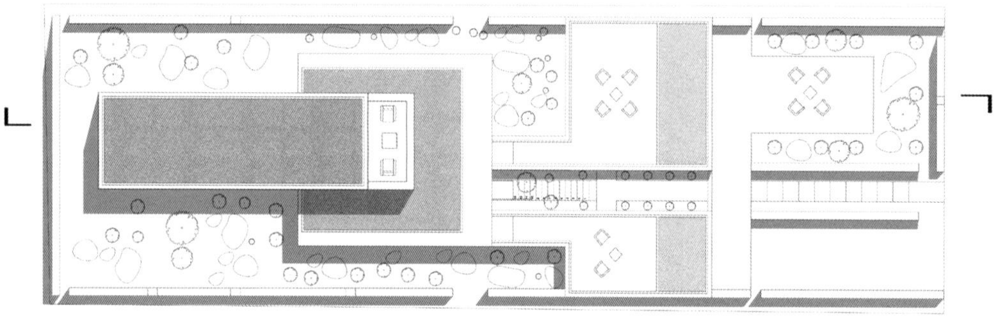

Roof plan

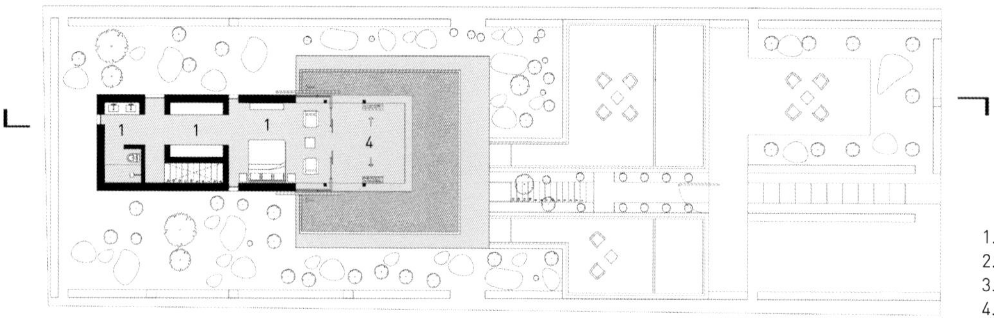

1. Bedroom
2. Dining room
3. Bathroom
4. Terrace

Third level plan

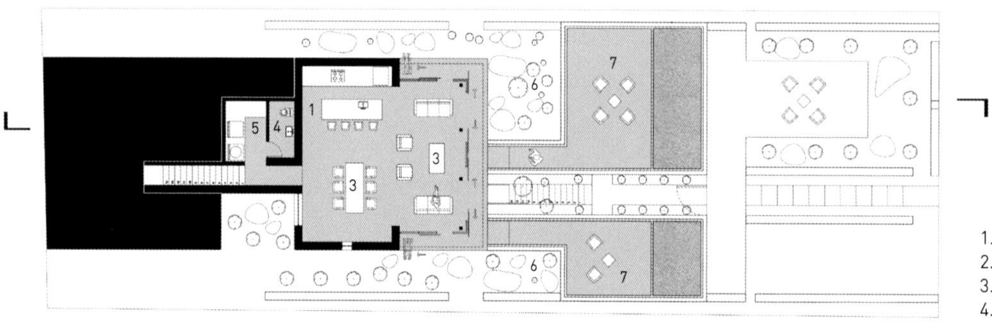

1. Kitchen
2. Dining room
3. Living room
4. Bathroom
5. Laundry
6. Garden
7. Terrace

Second level plan

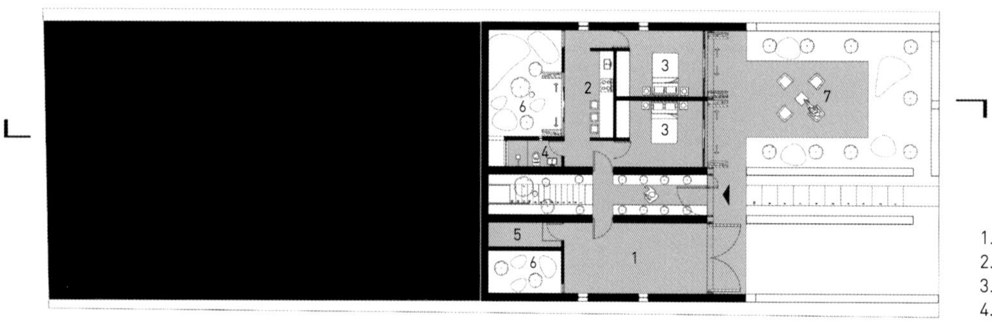

1. Parking
2. Kitchen
3. Bedroom
4. Bathroom
5. Storage
6. Garden
7. Terrace

First level plan

0 1 2 5 m

Prevalent, based in Sydney and working internationally, provides strategies for the built environment that prioritize environmental, economic, and social sustainability. With experience in both public and private sectors, they focus on user experience and one's relationship to the built environment at various scales. Prevalent's methodology is to utilize architecture to create a more inclusive, sustainable, and habitable built environment, ranging from large-scale projects to nano technologies. Their completed works include carbon neutral projects, funding material-based startups, dissolving city planning boundaries, and creating products that increase the health of internal spaces through the orchestration of light. Prevalent's intelligence and strategies behind each project set precedence within the industry.

Prevalent, mit Sitz in Sydney und international tätig, bietet Strategien für die gebaute Umwelt, die Umwelt-, Wirtschafts- und soziale Nachhaltigkeit priorisieren. Mit Erfahrung in öffentlichen und privaten Sektoren konzentrieren sie sich auf Benutzererfahrung und die Beziehung des Einzelnen zur gebauten Umwelt in verschiedenen Maßstäben. Die Methodologie von Prevalent besteht darin, Architektur zu nutzen, um eine inklusivere, nachhaltigere und bewohnbarere gebaute Umwelt zu schaffen, von Großprojekten bis hin zu Nanotechnologien. Ihre abgeschlossenen Arbeiten umfassen kohlenstoffneutrale Projekte, die Finanzierung von materialbasierten Start-ups, die Auflösung von Stadtplanungsgrenzen und die Schaffung von Produkten, die die Gesundheit von Innenräumen durch die Orchestrierung von Licht verbessern. Die Intelligenz und Strategien von Prevalent hinter jedem Projekt setzen Maßstäbe innerhalb der Branche.

MODUS OPERANDI BREWERY

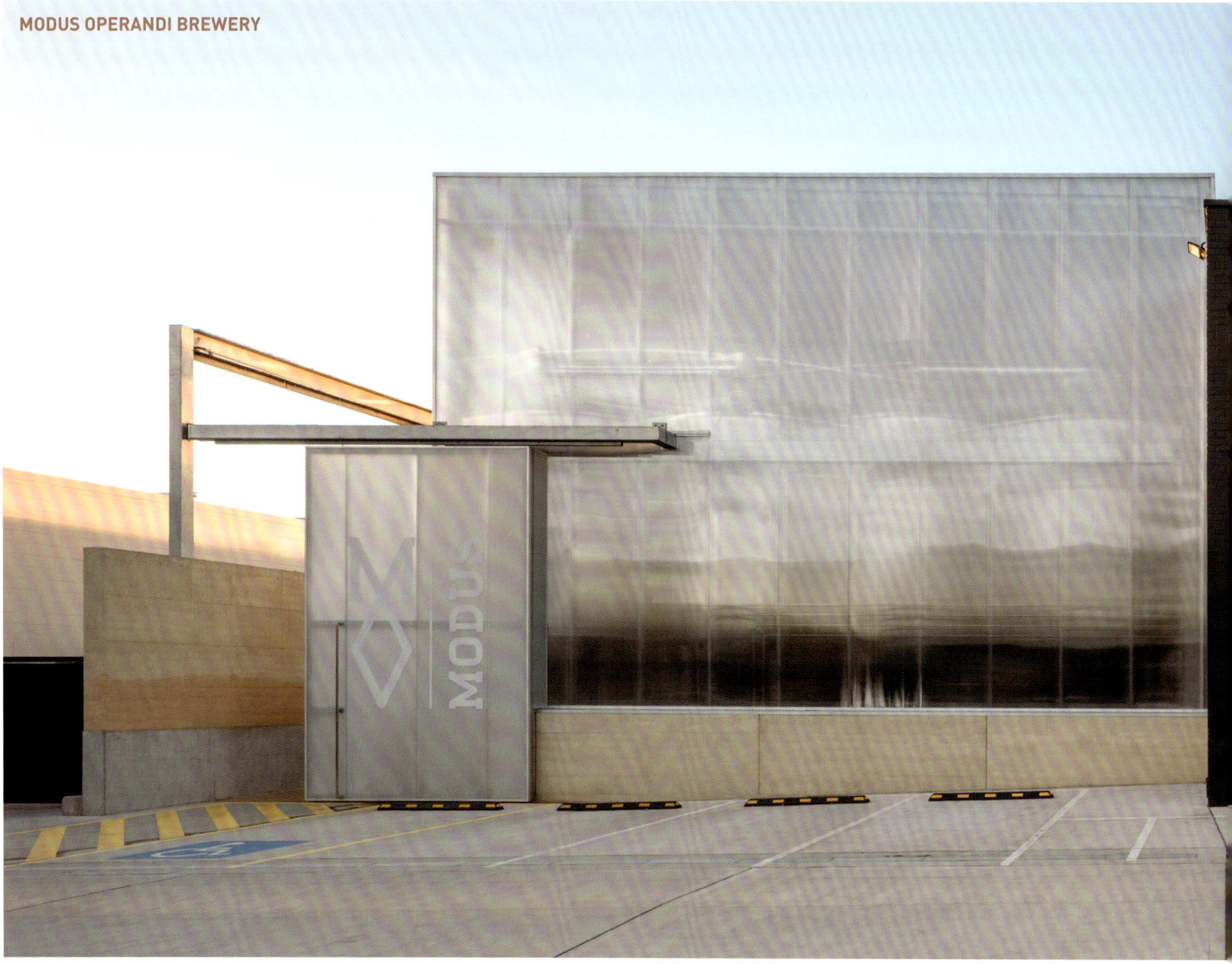

Prevalent, basé à Sydney et travaillant à l'international, fournit des stratégies pour l'environnement construit qui privilégient la durabilité environnementale, économique et sociale. Avec une expérience dans les secteurs public et privé, ils se concentrent sur l'expérience de l'utilisateur et sa relation avec l'environnement construit à différentes échelles. La méthodologie de Prevalent consiste à utiliser l'architecture pour créer un environnement construit plus inclusif, durable et habitable, allant des grands projets aux nanotechnologies. Leurs œuvres achevées comprennent des projets neutres en carbone, le financement de start-ups basées sur des matériaux, la dissolution des frontières de la planification urbaine et la création de produits qui améliorent la santé des espaces intérieurs grâce à l'orchestration de la lumière. L'intelligence et les stratégies de Prevalent derrière chaque projet établissent une prééminence dans l'industrie.

Prevalent, con sede en Sydney y trabajando a nivel internacional, proporciona estrategias para el entorno construido que priorizan la sostenibilidad ambiental, económica y social. Con experiencia en los sectores público y privado, se centran en la experiencia del usuario y la relación de este con el entorno construido a diferentes escalas. La metodología de Prevalent consiste en utilizar la arquitectura para crear un entorno construido más inclusivo, sostenible y habitable, abarcando desde proyectos a gran escala hasta nanotecnologías. Sus obras completadas incluyen proyectos neutros en carbono, financiación de startups basadas en materiales, disolución de límites de planificación urbana y creación de productos que aumentan la salud de los espacios internos a través de la orquestación de la luz. La inteligencia y las estrategias de Prevalent detrás de cada proyecto establecen un precedente dentro de la industria.

MODUS OPERANDI BREWERY

MEREWETHER, NSW AUSTRALIA // PHOTOS © JAN VRANOVSKÝ

Year: 2019-2021
Practice Team: Prevalent
Project Architect: Ben Berwick, Director; Joshua Healey, Student of Architecture; Sarah Veas, Graduate of Architecture
Builder: Kingston Building Australia
Town Planner: de Witt Consulting
Structural Engineer: Skelton Consulting Engineers
Landscape Consultant: Sustainable Surrounds
Hydraulic Consultant: McCallum PFCA
ESD Consultant: Evergreen Energy Consultants
Electrical Consultant: GDW Engineering
Civil Consultant: DRB Engineering
Rammed Earth Construction: Rammed Earth Australia
Rammed Earth Engineer: Hillbridge Engineers
Signage Installer: Hunter Signs
Signage Installer: Owen Signs
Upholsterer: Chapman Upholstery
Custom Fabricator: Defy Design
AV Consultant: Your Vision
Specialist Flooring Contractor: Roxset Australia
Access Consultant: Accessed
Fire Engineer: Design Confidence
Environmental Engineer: RCA Australia
Kitchen Consultant: Caterbuild
Stone Supplier: Central Coast Granite
Solid Surface: Hunter Joinery Group
Glazier: F&C
Joiner: Sydney Kitchens
Joiner: New Dimension
Refridgeration: Raisbeck Refridgeration
Fitout Construction: Ceil Construction

Our natural environment is in constant flux due to environmental degradation. This presents two positions in architecture: inflexible envelopes that protect between interior and exterior resulting in ever-fattening wall typologies or flexible envelopes whose orchestration is adapted by the user based on need. The Modus Operandi Brewery presents a case to the latter. The existing post-war portal frame shed is peeled open to one side, and a new sleeve-like addition added. Sliding planes wall 250 m^2 of space, allowing an external space to become internal and vice versa. The sliding mechanisms interface not only between interior and exterior but also between lot boundaries, blurring the limits of the site. The operability of the sliding partitions differs based on their size. The material palette is consistent throughout, with a sand floor and partly recycled polycarbonate planes. Sheds have allowed flexible interiors for various functions, Modus Operandi Brewery aiming to be an example of flexible architecture that reforms our relationship with the ever-changing landscape.

Unsere natürliche Umwelt befindet sich aufgrund von Umweltverschmutzung ständig im Wandel. Dies stellt in der Architektur zwei Positionen dar: Unflexible Hüllen, die zwischen Innen- und Außenraum schützen und zu immer fetteren Wandtypologien führen oder flexible Hüllen, deren Orchestrierung vom Benutzer je nach Bedarf angepasst wird. Die Modus Operandi Brewery stellt einen Fall für letzteres dar. Der bestehende Nachkriegsportalrahmen-Schuppen wird auf einer Seite aufgeschält und es wird eine neue ärmelartige Erweiterung hinzugefügt. Schiebeebenen umgeben 250 m^2 Raum, was einen externen Raum zu einem internen und umgekehrt werden lässt. Die Schiebemechanismen dienen nicht nur als Schnittstelle zwischen Innen- und Außenraum, sondern auch zwischen Grundstücksgrenzen und verschwimmen somit die Grenzen des Geländes. Die Betriebsfähigkeit der Schiebetrennwände unterscheidet sich je nach Größe. Die Materialpalette ist durchgehend, mit einem Sandboden und teilweise recycelten Polycarbonat-Ebenen. Schuppen haben flexible Innenräume für verschiedene Funktionen ermöglicht, und die Modus Operandi Brewery soll ein Beispiel für flexible Architektur sein, die unsere Beziehung zur sich ständig verändernden Landschaft reformiert.

Notre environnement naturel est en constante évolution en raison de la dégradation environnementale. Cela présente deux positions en architecture : des enveloppes rigides qui protègent entre l'intérieur et l'extérieur, ce qui entraîne des typologies de murs de plus en plus épais, ou des enveloppes flexibles dont l'orchestration est adaptée par l'utilisateur en fonction des besoins. La brasserie Modus Operandi présente un cas pour ce dernier. La remise existante à ossature en portail d'après-guerre est épluchée d'un côté et une nouvelle addition en forme de manche est ajoutée. Des murs de plans coulissants recouvrent 250 m^2 d'espace, permettant à un espace extérieur de devenir interne et vice versa. Les mécanismes coulissants interfèrent non seulement entre l'intérieur et l'extérieur, mais aussi entre les limites de terrain, brouillant les limites du site. L'opérabilité des cloisons coulissantes diffère en fonction de leur taille. La palette de matériaux est cohérente partout, avec un sol de sable et des plans de polycarbonate partiellement recyclé. Les remises ont permis des intérieurs flexibles pour diverses fonctions, et la brasserie Modus Operandi vise à être un exemple d'architecture flexible qui réforme notre relation avec le paysage en constante évolution.

Nuestro entorno natural está en constante cambio debido a la degradación ambiental. Esto presenta dos posiciones en la arquitectura: envolventes inflexibles que protegen entre el interior y el exterior, lo que resulta en tipologías de paredes cada vez más gruesas, o envolventes flexibles cuya orquestación se adapta por necesidad del usuario. La cervecería Modus Operandi presenta un caso para este último. El cobertizo existente con estructura de portal de posguerra se sitúa en un lado y se añade una nueva adición similar a una manga. Las paredes de planos deslizantes cubren 250 m^2 de espacio, permitiendo que un espacio exterior se convierta en interior y viceversa. Los mecanismos deslizantes no sólo interfieren entre el interior y el exterior, sino también entre los límites del terreno, difuminando los límites del lugar. La operabilidad de las particiones deslizantes difiere en función de su tamaño. La paleta de materiales es consistente en todas partes, con un suelo de arena y planos de policarbonato parcialmente reciclado. Los cobertizos han permitido interiores flexibles para varias funciones, con la cervecería Modus Operandi tiene como ejemplo de arquitectura flexible que reforma nuestra relación con el paisaje en constante cambio.

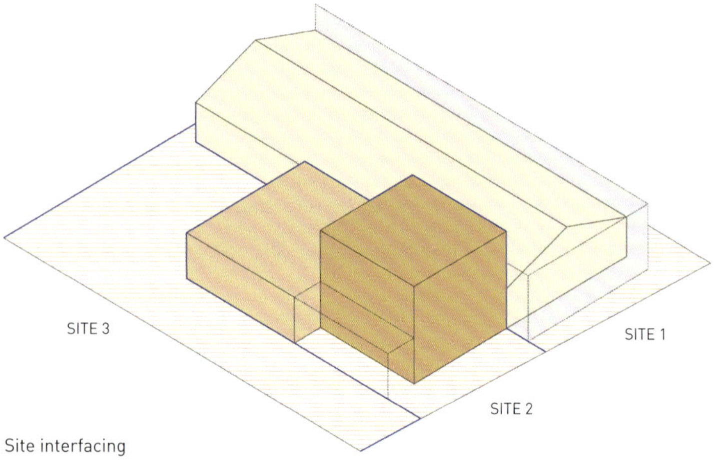

Site interfacing

Site 1: structure and form maintained
Site 2: interfacing elements on "blank site"
Site 3: future operations on "blank site"

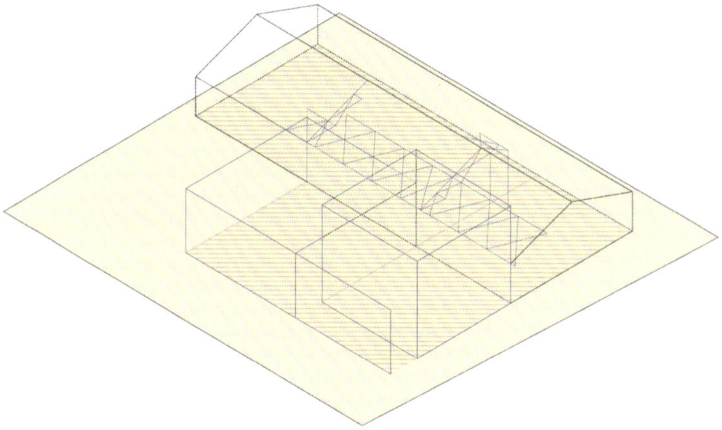

Structural stitching to allow free floorplan

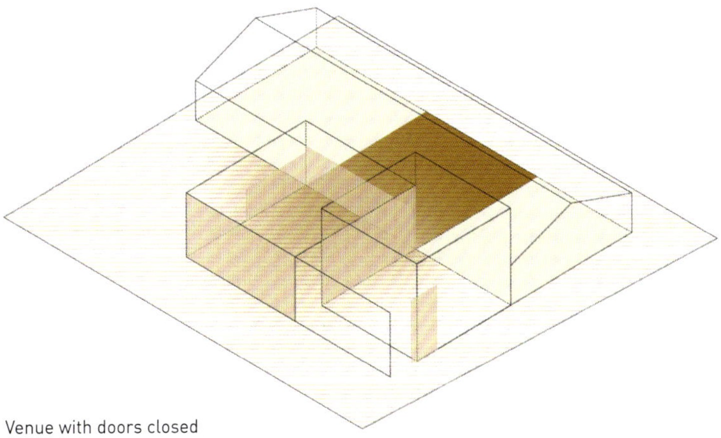

Venue with doors closed

Public

Production

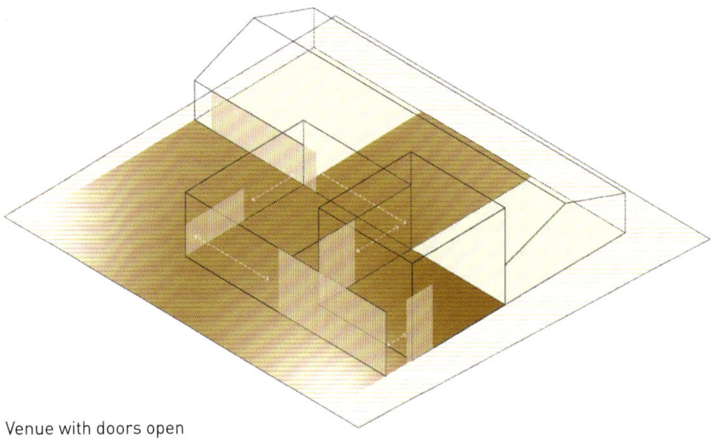

Venue with doors open

Public

Production

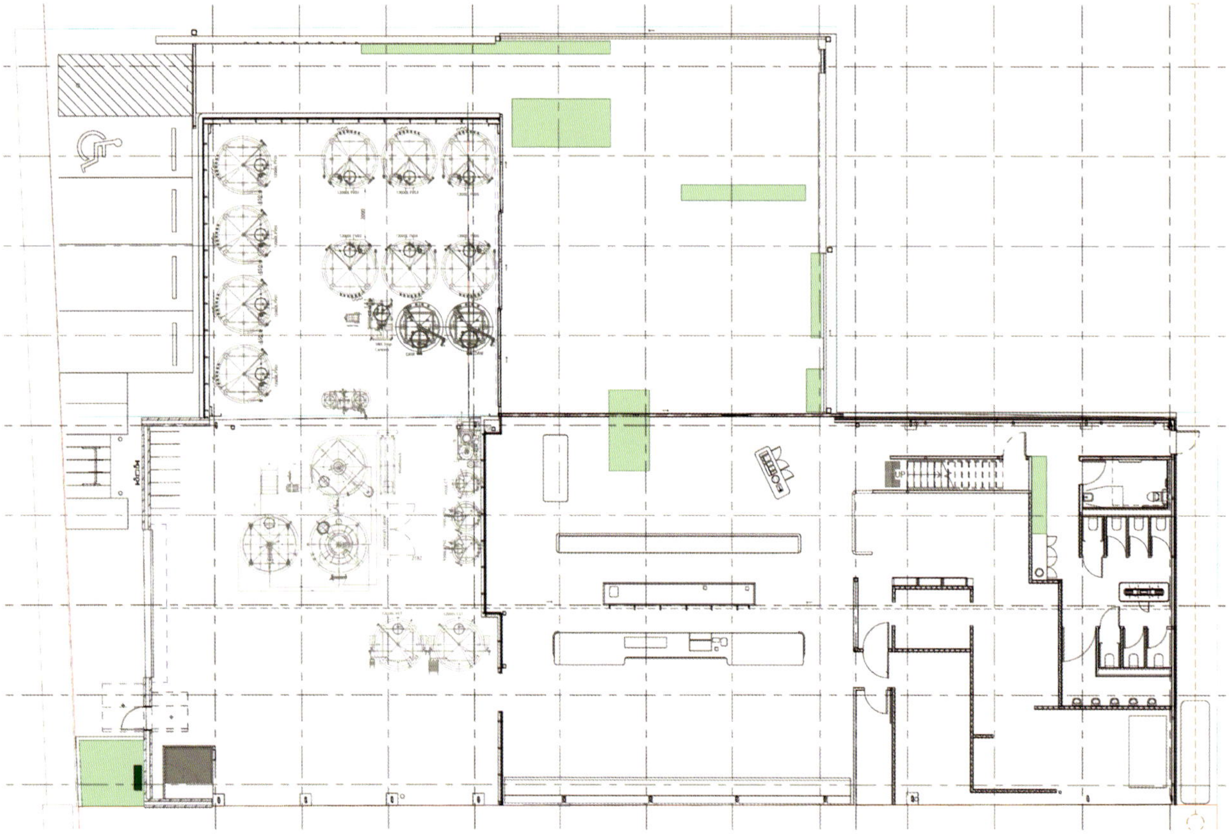

Floor plan

RINALDO DEL NERO

Architect-Engineer and CasaClima consultant, with a degree from Milan Polytechnic University, Rinaldo del Nero works in northern Italy within the alpine context of the Sondrio Province. He specialises in designing low environmental impact buildings and is particularly dedicated to renovation and adaptive reuse of historic rural structures in natural stone. Climate change, energy efficiency and environmental health are critical themes in his work, paired with scrupulous attention to architectural style, technical details, materiality and the surrounding context. The studio also designs bespoke homes, shops, offices, restaurants and hotels, creating unique experiential spaces.

RdN Studio employs a team of young professionals who collaborate with the goal of creating high quality, sustainable architecture that respects the surrounding territory.

Architekt-Ingenieur und CasaClima-Berater mit einem Abschluss der Polytechnischen Universität Mailand, Rinaldo del Nero arbeitet im Norden Italiens im alpinen Kontext der Provinz Sondrio. Er ist auf die Gestaltung von Gebäuden mit geringem Umwelteinfluss spezialisiert und widmet sich insbesondere der Renovierung und adaptiven Wiederverwendung historischer ländlicher Strukturen in Naturstein. Klimawandel, Energieeffizienz und Umweltgesundheit sind kritische Themen in seiner Arbeit, gepaart mit akribischer Aufmerksamkeit für architektonischen Stil, technische Details, Materialität und das umgebende Kontext. Das Studio entwirft auch maßgeschneiderte Wohnhäuser, Geschäfte, Büros, Restaurants und Hotels und schafft einzigartige Erlebnisräume.

RdN Studio beschäftigt ein Team junger Fachleute, die zusammenarbeiten, um hochwertige, nachhaltige Architektur zu schaffen, die das umgebende Territorium respektiert.

Architecte-ingénieur et consultant CasaClima, diplômé de l'Université Polytechnique de Milan, Rinaldo del Nero travaille dans le nord de l'Italie dans le contexte alpin de la province de Sondrio. Il est spécialisé dans la conception de bâtiments à faible impact environnemental et se consacre particulièrement à la rénovation et à la réutilisation adaptative de structures rurales historiques en pierre naturelle. Le changement climatique, l'efficacité énergétique et la santé environnementale sont des thèmes critiques dans son travail, associés à une attention scrupuleuse pour le style architectural, les détails techniques, la matérialité et le contexte environnant. Le studio conçoit également des maisons sur mesure, des magasins, des bureaux, des restaurants et des hôtels, créant des espaces d'expérience uniques.
RdN Studio emploie une équipe de jeunes professionnels qui collaborent dans le but de créer une architecture durable de haute qualité qui respecte le territoire environnant.

Arquitecto-Ingeniero y consultor de CasaClima, con un título de la Universidad Politécnica de Milán, Rinaldo del Nero trabaja en el norte de Italia en el contexto alpino de la provincia de Sondrio. Se especializa en el diseño de edificios de bajo impacto ambiental y está particularmente dedicado a la renovación y reutilización adaptativa de estructuras rurales históricas en piedra natural. El cambio climático, la eficiencia energética y la salud ambiental son temas críticos en su trabajo, combinados con una atención escrupulosa al estilo arquitectónico, los detalles técnicos, la materialidad y el contexto circundante. El estudio también diseña hogares personalizados, tiendas, oficinas, restaurantes y hoteles, creando espacios de experiencia únicos.
RdN Studio emplea a un equipo de jóvenes profesionales que colaboran con el objetivo de crear una arquitectura sostenible de alta calidad que respeta el territorio circundante.

TERRA DEL SOLE SUITES

ALBOSAGGIA, VALTELLINA, ITALY // PHOTOS © MARCELLO MARIANA

Project: Rinaldo Del Nero
Concept Design: Anna Barlascini
Consultants: Skye Sturm, Cristiano Del Nero, Andrea Mainetti
Producers: Karman, Rexa, Tom Dixon, Florim, Kerakoll, Claytec
Furnishing and Windows: Bricalli FBF
Plasterboard: 3 Botta srl
Construction: CLV costruzioni
Painting: Morella snc
Flooring: Spandrio srl
Lighting: E27 illuminazione
Electric: Ciapponi Impianti
Plumbing: GMT Giugni
Drapery: Tecnotenda Sciolini
Resin: Resina Valtellina
Building Supplies: Biomarket srl

Terra del Sole Agriturismo is a charming rural retreat in Albosaggia, Valtellina, situated amidst stunning natural surroundings including the Adda River, lush fields, and picturesque alpine peaks. The facility comprises two suites, each with its own unique design, inspired by nature. In Suite Terra, the use of rammed earth walls built entirely by hand with no chemical bonding agents creates a beautiful, natural room divider that also serves as a "green" lung. The walls purify the air, regulate humidity, and promote a sense of harmony and health in the room. The decor features furniture built from natural wood, evocative organic light fixtures, and flooring reminiscent of the vineyard soil. The room offers stunning views of the surrounding alpine landscape and provides a serene space for guests to reconnect with nature and themselves.

Terra del Sole Agriturismo ist ein bezauberndes Landhotel in Albosaggia, Valtellina, das sich inmitten atemberaubender natürlicher Umgebung befindet, einschließlich des Flusses Adda, üppiger Felder und malerischer alpiner Gipfel. Die Anlage besteht aus zwei Suiten, von denen jede über ein einzigartiges Design verfügt, das von der Natur inspiriert ist. In der Suite Terra wird durch die Verwendung von mit bloßen Händen vollständig aus Stampflehm gebauten Wänden ohne chemische Bindemittel ein schöner natürlicher Raumteiler geschaffen, der auch als „grüne Lunge" dient. Die Wände reinigen die Luft, regulieren die Feuchtigkeit und fördern ein Gefühl von Harmonie und Gesundheit im Raum. Die Einrichtung besteht aus Möbeln aus Naturholz, evokativen organischen Leuchten und Böden, die an den Boden in Weinbergen erinnern. Das Zimmer bietet einen atemberaubenden Blick auf die umliegende alpine Landschaft und bietet einen ruhigen Raum, um sich mit der Natur und sich selbst zu verbinden.

Terra del Sole Agriturismo est une charmante retraite rurale à Albosaggia, Valtellina, située au milieu d'un environnement naturel époustouflant comprenant la rivière Adda, des champs luxuriants et des pics alpins pittoresques. L'établissement comprend deux suites, chacune avec son propre design unique, inspiré de la nature. Dans la Suite Terra, l'utilisation de murs en terre battue construits entièrement à la main sans agents de liaison chimiques crée un magnifique séparateur de pièce naturel qui sert également de « poumon vert ». Les murs purifient l'air, régulent l'humidité et favorisent un sentiment d'harmonie et de santé dans la pièce. La décoration comprend des meubles en bois naturel, des luminaires organiques évocateurs et un sol rappelant le sol des vignobles. La chambre offre une vue imprenable sur le paysage alpin environnant et offre un espace serein pour les clients pour se reconnecter avec la nature et eux-mêmes.

Terra del Sole Agriturismo es un encantador retiro rural en Albosaggia, Valtellina, situado en un entorno natural impresionante que incluye el río Adda, campos exuberantes y pintorescas cumbres alpinas. La instalación consta de dos suites, cada una con su propio diseño único inspirado en la naturaleza. En Suite Terra, el uso de paredes de tierra apisonada construidas completamente a mano sin agentes químicos de unión crea un hermoso divisor de habitación natural que también sirve como «pulmón verde». Las paredes purifican el aire, regulan la humedad y promueven una sensación de armonía y salud en la habitación. La decoración cuenta con muebles construidos con madera natural, evocadoras luminarias orgánicas y un piso que recuerda al suelo de los viñedos. La habitación ofrece impresionantes vistas del paisaje alpino circundante y proporciona un espacio sereno para que los huéspedes se reconecten con la naturaleza y consigo mismos.

North wall

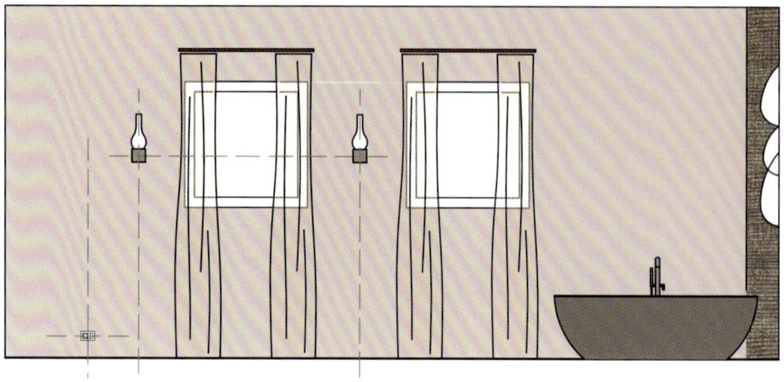
South wall

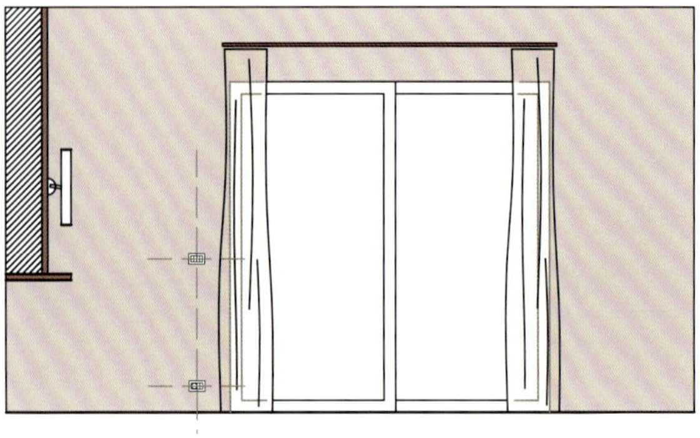
East wall

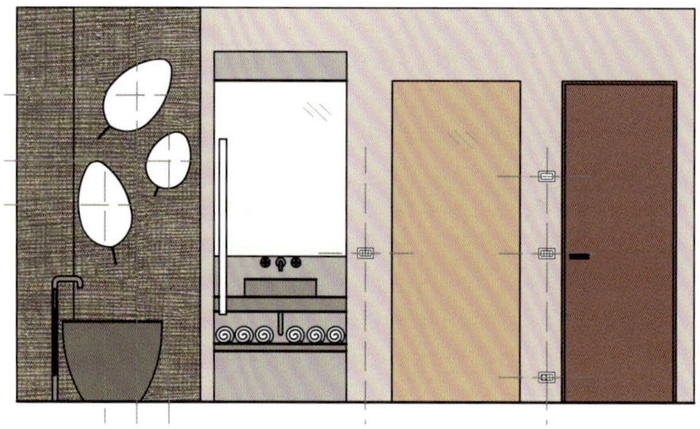
West wall

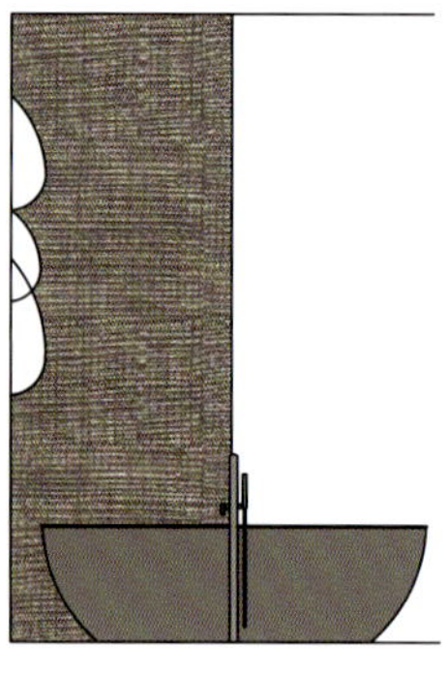
North wall (tub area)

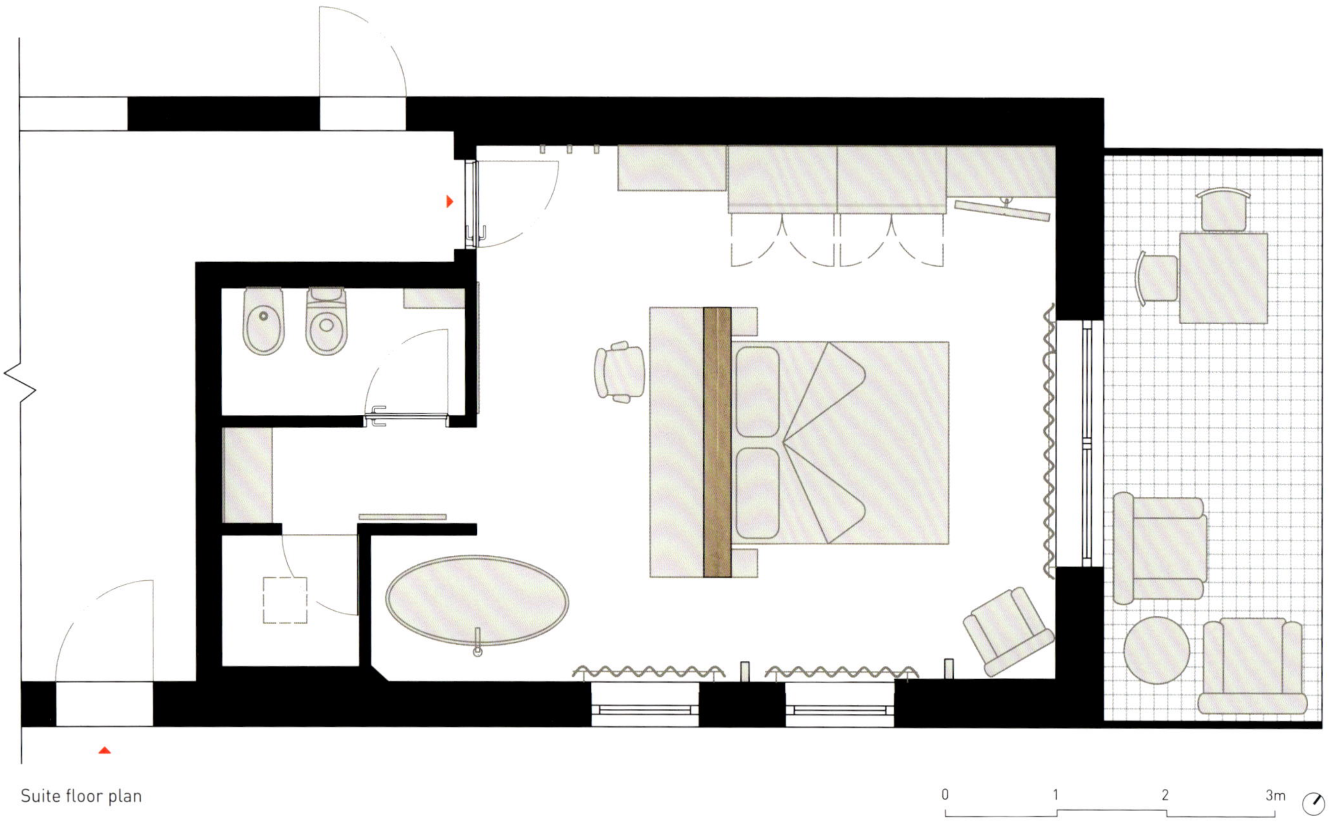

Suite floor plan

Illustration © Paolo Barlascini

Sfera Blu was founded in 2007 in Ahmedabad with an idea to question the cookie cutter approach in design. This approach led to the creation of few projects which were quite experimental in nature. Slowly a method was created which involved being introduced with new materials on a regular basis, figuring out its characteristics and giving it a new personality in our buildings. This process has led us to collaborate with some really amazing minds in the creative space. In the last 15 years we have learnt that trees, insects, animals and birds should be considered in the same category as walls and columns in respect to architecture. Nature plays an important role in our design. We are in the middle of integrating future technology to analyze how much integration with surrounding environment is essential.

Sfera Blu wurde 2007 in Ahmedabad mit der Idee gegründet, den Cookie-Cutter-Ansatz im Design zu hinterfragen. Dieser Ansatz führte zur Entwicklung einiger Projekte, die eher experimenteller Natur waren. Nach und nach wurde eine Methode entwickelt, bei der wir regelmäßig neue Materialien kennenlernen, ihre Eigenschaften herausfinden und ihnen in unseren Gebäuden eine neue Persönlichkeit verleihen. Dieser Prozess hat uns dazu gebracht, mit einigen wirklich erstaunlichen Köpfen im kreativen Bereich zusammenzuarbeiten. In den letzten 15 Jahren haben wir gelernt, dass Bäume, Insekten, Tiere und Vögel in Bezug auf die Architektur in der gleichen Kategorie wie Wände und Säulen betrachtet werden sollten. Die Natur spielt in unserem Design eine wichtige Rolle. Wir sind gerade dabei, zukünftige Technologien zu integrieren, um zu analysieren, inwieweit die Integration mit der Umgebung wichtig ist.

Sfera Blu a été fondée en 2007 à Ahmedabad avec l'idée de remettre en question l'approche à l'emporte-pièce du design. Cette approche a conduit à la création de quelques projets de nature expérimentale. Peu à peu, une méthode s'est mise en place, qui consiste à se familiariser régulièrement avec de nouveaux matériaux, à comprendre leurs caractéristiques et à leur donner une nouvelle personnalité dans nos bâtiments. Ce processus nous a amenés à collaborer avec des esprits vraiment extraordinaires dans le domaine de la création. Au cours des 15 dernières années, nous avons appris que les arbres, les insectes, les animaux et les oiseaux devraient être considérés dans la même catégorie que les murs et les colonnes en ce qui concerne l'architecture. La nature joue un rôle important dans notre conception. Nous sommes en train d'intégrer les technologies du futur afin d'analyser dans quelle mesure l'intégration avec le milieu environnant est essentielle.

Sfera Blu se fundó en 2007 en Ahmedabad con la idea de cuestionar el enfoque del diseño basado en un molde. Este enfoque llevó a la creación de algunos proyectos de carácter experimental. Poco a poco se fue creando un método que consistía en introducir nuevos materiales con regularidad, averiguar sus características y darles una nueva personalidad en nuestros edificios. Este proceso nos ha llevado a colaborar con algunas mentes realmente asombrosas en el espacio creativo. En los últimos 15 años hemos aprendido que los árboles, los insectos, los animales y los pájaros deben considerarse de la misma categoría que los muros y las columnas en lo que respecta a la arquitectura. La naturaleza desempeña un papel importante en nuestro diseño. Estamos en pleno proceso de integración de la tecnología del futuro para analizar hasta qué punto es esencial la integración con el entorno.

MEETHI-MISHTI NU MATI GHAR

AHMEDABAD, GUJARAT, INDIA // PHOTOS © UMANG SHAH

Social Media Accounts: www.instagram.com/sferabluarchitects/
www.instagram.com/naman2312/
www.facebook.com/blusphere5
Contact: sferablu@gmail.com
Firm Location: Ahmedabad, Gujarat, India
Completion Year: 2020
Gross Built Area: 440 m^2
Program: Residence for 2 children aged 3 & 9. This house has a Parent Bedroom, Children Bedroom and a Guest Bedroom. A large Living Room with a porch opening towards the garden. Kitchen and Dining room. A Library, Secret Room and A Late Night TV viewing Room. Supporting spaces such as Toilets and Stores.
Lead Architect: Naman Shah
Lead Architects e-mail: naman2312@gmail.com

Meethi and Mishti were clients, aged 3 and 9, respectively, and wanted an eco-friendly house that would not harm the nature around them. They suggested using minimal cement and bringing in less material from elsewhere, so the earth from the site was used to build the house through the rammed earth technology, which uses only 6-8 % of cement normally required for construction. Different natural oxides were used to create layered fluid patterns on the walls, and the girls' stone and shell collections were added to preserve their fond memories. Upcycled wood from a demolished building was used to create frames, doors, and windows. A dynamic ceiling made of sloping glass was designed to let diffused north light flow through the house, giving it a cool and airy feel. The house also included a playhouse with a slide, a secret room with three different entries, an acrobatic bed in the girls' room, and a bridge that ramps down to the girl's room for an impromptu catwalk. Blackboards were used as partitions, and a hidden basin was designed to address the girls' concern about seeing it while eating. Sketches from the girls' drawing books were used to create grills for the windows. The house embodies the people who live in it, and is both humane and honest.

Meethi und Mishti, 3 und 9 Jahre alt, wollten ein umweltfreundliches Haus, das die Natur um sie herum schützt. Das Haus wurde durch Stampflehmbau errichtet, der nur 6-8 % des normalerweise verwendeten Zements benötigt. Natürliche Oxide wurden verwendet, um Schichtflüssigkeitsmuster auf den Wänden zu schaffen. Lokale Materialien mit niedrigem Kohlenstoff-Fußabdruck wurden verwendet und aufgearbeitetes Holz aus einem abgerissenen Gebäude wurde für Türen und Fenster verwendet. Kalkputz wurde verwendet, um Toxine aus der Luft zu filtern, und eine Decke aus schrägem Glas wurde entworfen, um diffuses Nordlicht durch das Haus fließen zu lassen. Das Haus hatte ein Spielhaus mit Rutsche, ein Geheimzimmer mit drei Eingängen, ein akrobatisches Bett und eine Brücke, die als improvisierter Laufsteg diente. Tafeln wurden als Trennwände verwendet, und ein verstecktes Waschbecken wurde entworfen, um es beim Essen zu verbergen. Das Haus verkörpert seine Bewohner und ist menschlich und ehrlich.

Meethi et Mishti, âgées de 3 et 9 ans respectivement, souhaitaient une maison écologique qui préserverait la nature environnante. Elles ont proposé d'utiliser peu de ciment et de favoriser les matériaux locaux, utilisant ainsi la terre du site pour construire la maison selon la technique de terre battue, limitant l'utilisation de ciment à 6-8 %. Les murs ont été ornés de motifs en oxydes naturels et décorés avec les collections de pierres et de coquillages des filles. Le bois recyclé d'un bâtiment démoli a été utilisé pour les cadres, les portes et les fenêtres, minimisant l'empreinte carbone. Le plafond dynamique incliné en verre a été conçu pour diffuser la lumière nordique et donner une sensation fraîche et aérée. La maison incluait également une maisonnette de jeux avec toboggan, une pièce secrète à trois entrées, un lit acrobatique dans la chambre des filles et un pont en pente pour une passerelle improvisée. Les cloisons étaient en ardoise et un lavabo discret a été créé pour répondre à la demande des filles de ne pas le voir pendant les repas. Les dessins des filles ont été utilisés pour créer des grilles pour les fenêtres. La maison reflète ses habitantes et est à la fois humaine et honnête.

Meethi y Mishti, clientes de 3 y 9 años, querian una casa ecológica que no dañara la naturaleza local. Propusieron usar el mínimo de cemento y traer menos materiales de otras partes, usando la tierra del lugar para construir la casa con tecnología de tierra apisonada, que solo usa el 6-8 % del cemento requerido normalmente. Se usaron óxidos naturales para crear patrones en las paredes y las colecciones de piedras y conchas de las niñas se agregaron para preservar sus recuerdos. Se usaron materiales locales de bajo impacto ambiental y madera reutilizada para marcos, puertas y ventanas. Un techo de vidrio inclinado deja pasar la luz difusa del norte, dándole una sensación fresca y aireada. También incluye una casita de juegos con tobogán, habitación secreta, cama acrobática y un puente que baja a la habitación de las niñas como una pasarela improvisada. Los bocetos de las niñas se usaron para crear rejillas para las ventanas. La casa encarna a las personas que viven en ella, siendo humana y honesta.

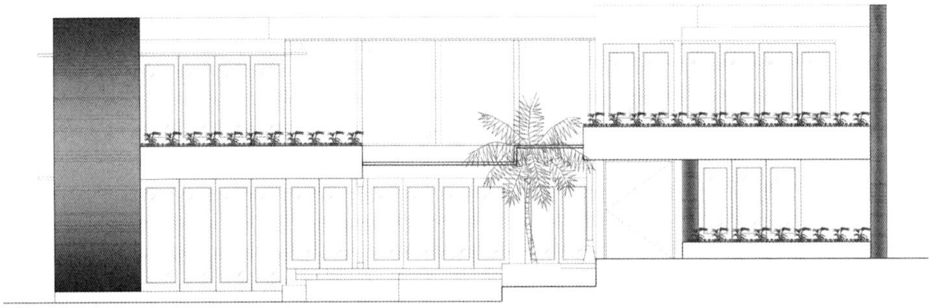

North elevation

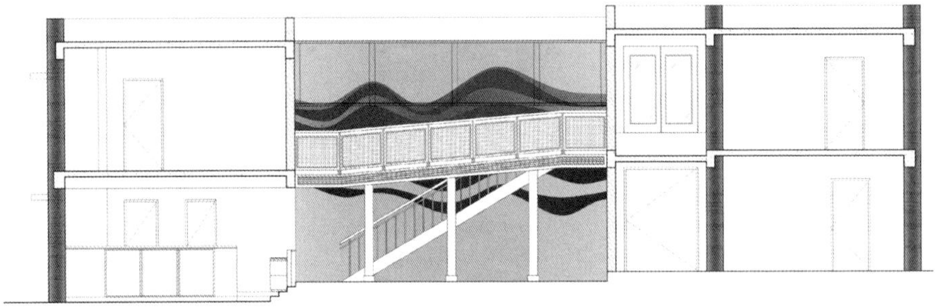

Section A-A'

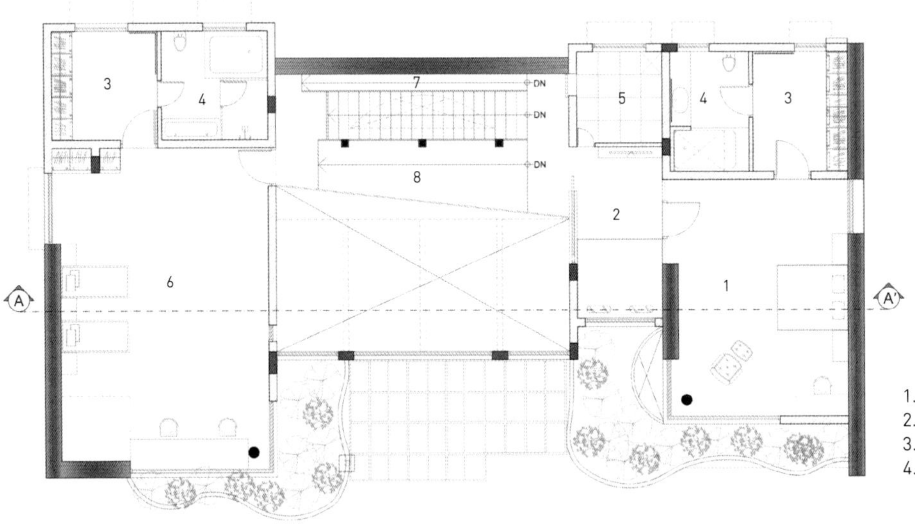

1. Master bedroom
2. TV room
3. Dressing
4. Toilet
5. Secret room
6. Daughter's room
7. Slide
8. Ramp

First floor plan

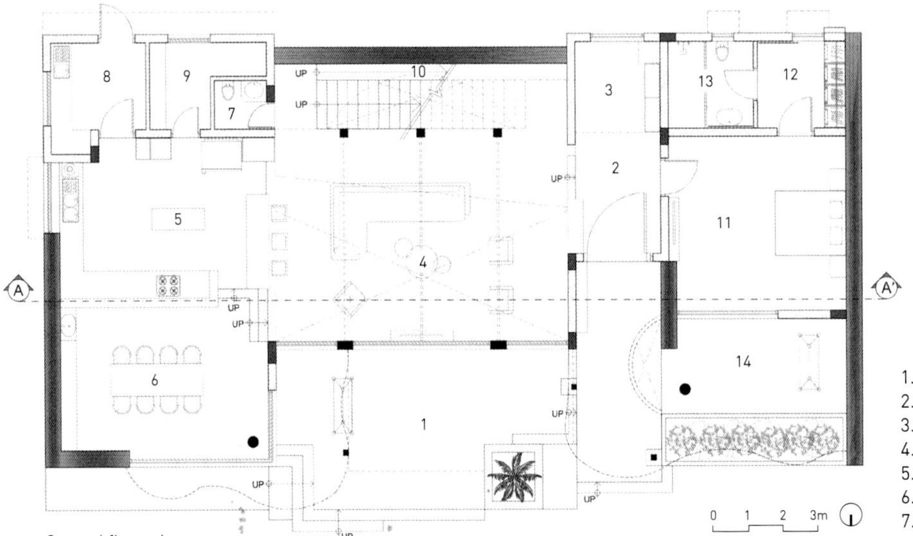

1. Patio
2. Entrance lobby
3. Library
4. Living room
5. Kitchen
6. Dining room
7. Common toilet
8. Utility
9. Store
10. Slide
11. Guest bedroom
12. Dressing
13. Toilet
14. Veranda

Ground floor plan

From a base in Bali, Indonesia, Studio Nicholas Burns and Workshop operate internationally throughout Asia, Europe and Africa. Nicholas Burns, Australian by birth established the studio after leaving university under the guidance of Professor Michael Tawa and has since been influenced and mentored by Richard Leplastrier, Glenn Murcutt and Peter Stutchbury. The studios' work seeks to define deep connections with place and landscape. Creating spaces of moving experience, touching emotions and drawing one closer to nature. Placing the sensory quality as paramount, beyond the materiality. Material authenticity and close collaboration with consultants result in a coherent whole without functional sacrifices or the deployment of irrelevant devices. The design intention is a sense of quiet both inside and out. The Studios' work continues to learn from different cultures and climates, materials, places and technologies to further develop and deepen the design language and understanding. To date the Studios have carried out work in Australia, Singapore, Indonesia, Malaysia, Thailand, Japan, Hong Kong, South Korea, New Zealand, Portugal and São Tomé and Príncipe.

Vom Standort Bali, Indonesien, aus arbeiten das Studio Nicholas Burns und Workshop international in ganz Asien, Europa und Afrika. Nicholas Burns, gebürtiger Australier, gründete das Studio nach seinem Abschluss an der Universität unter der Anleitung von Professor Michael Tawa und wurde seitdem von Richard Leplastrier, Glenn Murcutt und Peter Stutchbury beeinflusst und unterstützt. Die Arbeit des Studios zielt darauf ab, tiefe Verbindungen zum Ort und zur Landschaft herzustellen. Es werden Räume geschaffen, die bewegende Erfahrungen bieten, Emotionen berühren und einen näher zur Natur bringen. Dabei steht die sinnliche Qualität im Vordergrund, über die Materialität hinaus. Eine Materialauthentizität und enge Zusammenarbeit mit Beratern führen zu einem schlüssigen Ganzen ohne funktionale Einschränkungen oder den Einsatz irrelevanter Geräte. Die Designintention ist eine ruhige Atmosphäre sowohl innen als auch außen. Das Studio setzt seine Arbeit fort, indem es von verschiedenen Kulturen und Klimazonen, Materialien, Orten und Technologien lernt, um die Designsprache und das Verständnis weiterzuentwickeln und zu vertiefen. Bisher hat das Studio in Australien, Singapur, Indonesien, Malaysia, Thailand, Japan, Hongkong, Südkorea, Neuseeland, Portugal und São Tomé und Príncipe gearbeitet.

AIREYS HOUSE

À partir de Bali, en Indonésie, le Studio Nicholas Burns et Workshop opère à l'international en Asie, en Europe et en Afrique. Nicholas Burns, Australien de naissance, a créé le studio après avoir quitté l'université sous la direction du professeur Michael Tawa et a depuis été influencé et mentoré par Richard Leplastrier, Glenn Murcutt et Peter Stutchbury. Le travail du studio vise à définir des liens profonds avec le lieu et le paysage. En créant des espaces d'expérience en mouvement, touchant les émotions et rapprochant de la nature. Plaçant la qualité sensorielle en priorité, au-delà de la matérialité. L'authenticité des matériaux et la collaboration étroite avec les consultants aboutissent à un tout cohérent sans sacrifices fonctionnels ni déploiement d'appareils inutiles. L'intention de conception est un sentiment de calme à l'intérieur et à l'extérieur. Le travail du studio continue d'apprendre des différentes cultures et climats, des matériaux, des lieux et des technologies pour développer et approfondir le langage et la compréhension de la conception. À ce jour, le studio a travaillé en Australie, à Singapour, en Indonésie, en Malaisie, en Thaïlande, au Japon, à Hong Kong, en Corée du Sud, en Nouvelle-Zélande, au Portugal et à São Tomé et Príncipe.

Desde su base en Bali, Indonesia, Studio Nicholas Burns y Workshop operan internacionalmente en toda Asia, Europa y África. Nicholas Burns, australiano de nacimiento, estableció el estudio después de dejar la universidad bajo la guía del profesor Michael Tawa y desde entonces ha sido influenciado y mentorizado por Richard Leplastrier, Glenn Murcutt y Peter Stutchbury. El trabajo del estudio busca definir conexiones profundas con el lugar y el paisaje. Creando espacios de experiencia en movimiento, que toquen las emociones y acerquen a uno a la naturaleza. Colocando la calidad sensorial como lo más importante, más allá de la materialidad. La autenticidad del material y la estrecha colaboración con consultores resultan en un todo coherente sin sacrificios funcionales o el despliegue de dispositivos irrelevantes. La intención de diseño es una sensación de tranquilidad tanto dentro como fuera. El trabajo del estudio continúa aprendiendo de diferentes culturas y climas, materiales, lugares y tecnologías para desarrollar y profundizar el lenguaje y la comprensión del diseño. Hasta la fecha, el estudio ha llevado a cabo trabajos en Australia, Singapur, Indonesia, Malasia, Tailandia, Japón, Hong Kong, Corea del Sur, Nueva Zelanda, Portugal y São Tomé y Príncipe.

AIREYS HOUSE

VICTORIA, AUSTRALIA // PHOTOS © PETER BENNETTS

Design Architect: Nicholas Burns
Client Project Managers: N.A.
The Collaborative Artwork by: N.A.
Landscape Architect: N.A.
Civil Engineer: Engineering Elements, Melbourne

The house in Aireys Inlet, Australia is located on a wooded site and has views of the ocean, lighthouse, and valley. The design of the house fits the site and existing vegetation, with rammed earth walls framing the view and connecting the experience to nature. The east-west orientation allows winter sun to penetrate the living areas and cool southern breezes to flow in from the ocean. Framed windows and openings allow for natural ventilation and cross-flow of air, while intimate niches provide views for quiet reflection. The bedrooms are protected yet offer panoramic views of the landscape. The design is attuned to the rhythms of nature and the environment, seeking a deep connection with nature.

Das Haus in Aireys Inlet, Australien, befindet sich auf einem bewaldeten Grundstück und bietet Aussicht auf den Ozean, den Leuchtturm und das Tal. Das Design des Hauses passt sich an den Standort und die vorhandene Vegetation an, mit Wänden aus gestampfter Erde, die die Aussicht umrahmen und das Erlebnis mit der Natur verbinden. Die Ost-West-Ausrichtung ermöglicht es der Wintersonne, in die Wohnbereiche zu gelangen, und kühle Brisen vom Süden aus dem Ozean zu fangen. Eingefasste Fenster und Öffnungen ermöglichen natürliche Belüftung und Querlüftung, während intime Nischen Ansichten für ruhige Reflexion bieten. Die Zimmer sind geschützt und bieten gleichzeitig einen Panoramablick auf die Landschaft. Das Design passt sich den Rhythmen der Natur und der Umgebung an und sucht eine tiefe Verbindung zur Natur.

La maison à Aireys Inlet, Australie, est située sur un terrain boisé avec vue sur l'océan, le phare et la vallée. La conception de la maison s'adapte au site et à la végétation existante, avec des murs en terre battue encadrant la vue et reliant l'expérience à la nature. L'orientation est-ouest permet au soleil d'hiver de pénétrer dans les espaces de vie et de capturer les brises fraîches du sud provenant de l'océan. Les fenêtres encadrées et les ouvertures permettent une ventilation naturelle et un flux d'air croisé, tandis que les niches intimes offrent des vues pour la réflexion tranquille. Les chambres sont protégées et offrent en même temps une vue panoramique sur le paysage. La conception s'adapte aux rythmes de la nature et de l'environnement, cherchant une connexion profonde avec la nature.

La casa en Aireys Inlet, Australia, está ubicada en un terreno boscoso y cuenta con vistas al océano, al faro y al valle. El diseño de la casa se ajusta al sitio y la vegetación existente, con muros de tierra apisonada que enmarcan la vista y conectan la experiencia con la naturaleza. La orientación este-oeste permite que el sol de invierno penetre en las áreas de estar y que las brisas frescas del sur lleguen desde el océano. Las ventanas enmarcadas y las aberturas permiten la ventilación natural y el flujo de aire cruzado, mientras que los nichos íntimos ofrecen vistas para la reflexión tranquila. Las habitaciones están protegidas y, al mismo tiempo, proporcionan una vista panorámica del paisaje. El diseño se ajusta a los ritmos de la naturaleza y el entorno, y busca una conexión profunda con la naturaleza.

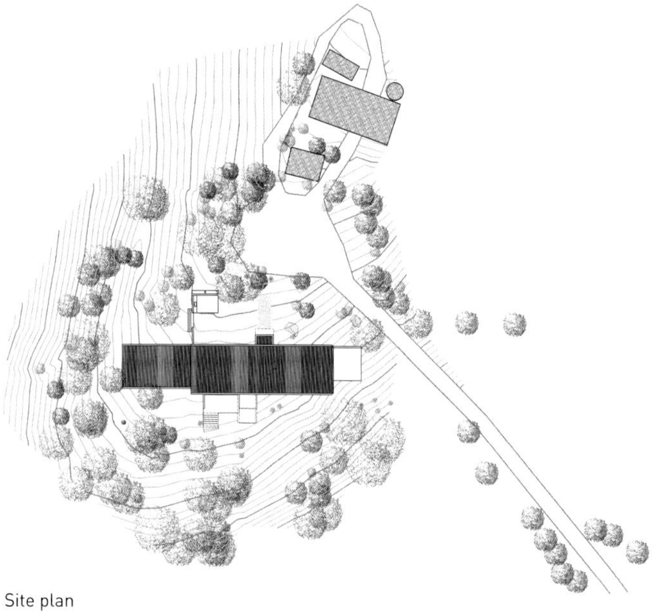

Site plan

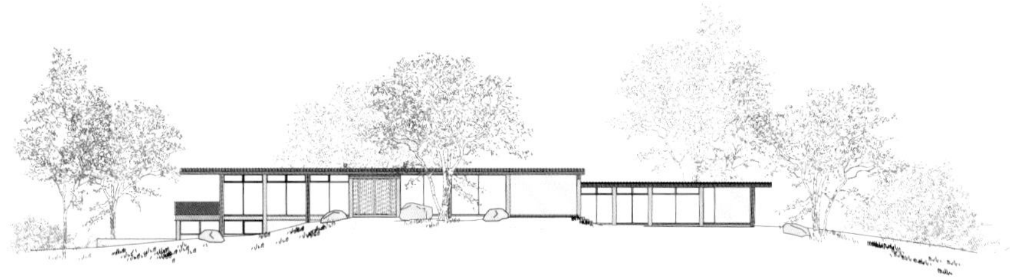

North elevation

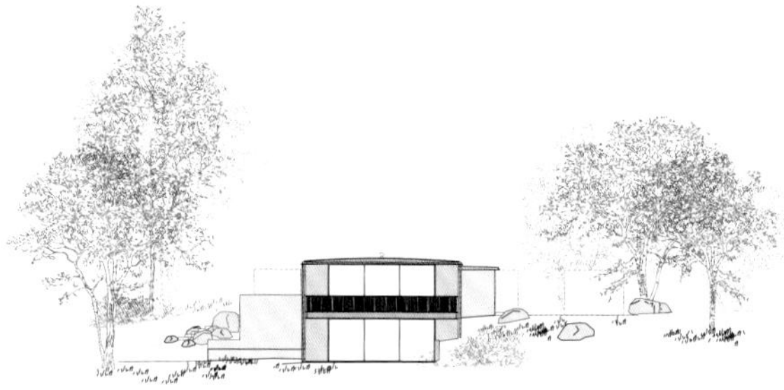

East elevation

South elevation

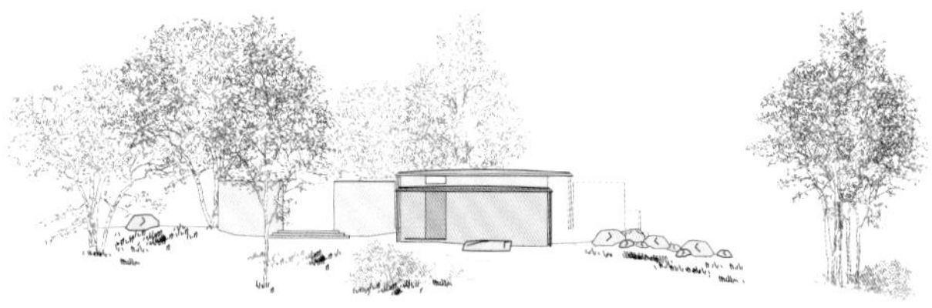

West elevation

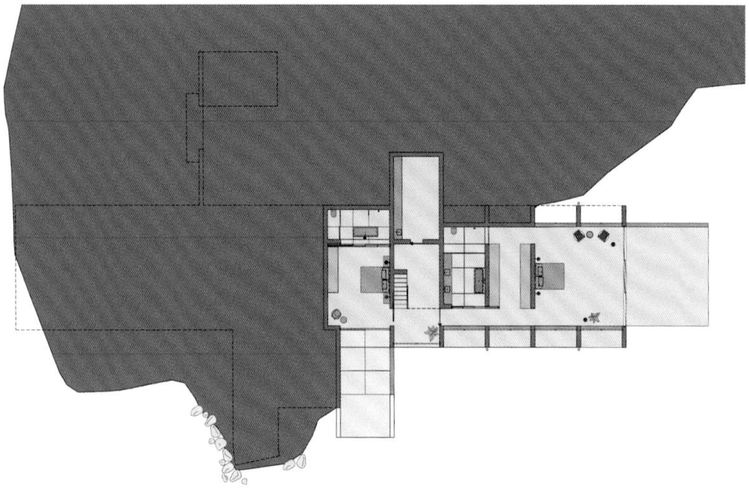

Lower ground floor plan

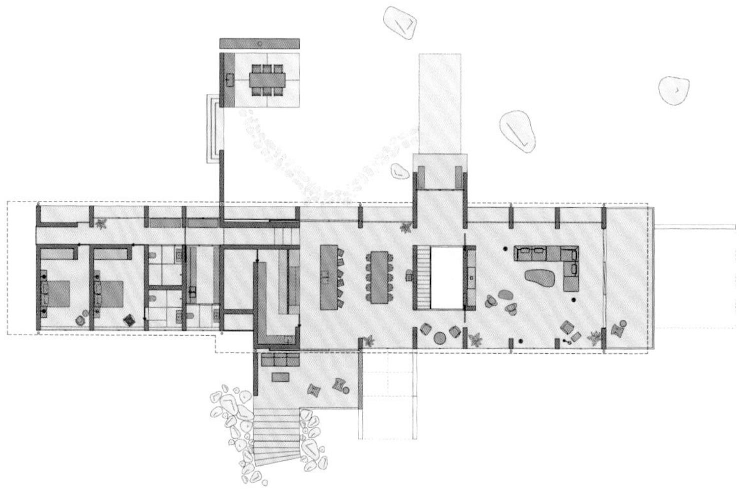

Ground floor plan

JULIMAR FALLS HOUSE

PERTH, AUSTRALIA // PHOTOS © PETER BENNETTS

Design Architect: Nicholas Burns
Client Project Managers: N.A.
The Collaborative Artwork by: N.A.
Landscape Architect: N.A.
Civil Engineer: Roger Pateman
Materials: rammed earth walls, concrete screed floors, steel / galvanised iron roof, stone - solid carved basins and bath, marri timber

Located in the Chittering valley, one and a half hours north of Perth, the landscape is a mix of cleared farming areas and bushland. The house is nestled into a steeply sloped hill, creating a refuge and a frame for connecting occupants to the environment. The design provides a series of interrelated experiences using low maintenance, weathering materials, leaving an unfinished look that integrates with the landscape. Open living spaces connect with courtyards and decking areas, framing nature and providing contrasting outdoor spaces for sitting and eating. Rammed earth walls provide thermal mass, and high-performance glazing, ventilation, and geothermal convective systems maintain comfort with low energy consumption. The house serves as a private retreat for the owners and a meeting place for extended family, with a courtyard garden in raised corten beds for producing fruit, herbs, and vegetables. Museum-grade LED lighting displays art to its best and provides essential quality without natural light.

Gelegen im Chittering-Tal, eineinhalb Stunden nördlich von Perth, besteht die Landschaft aus einer Mischung von gerodeten Ackerflächen und Buschland. Das Haus ist in einen steil abfallenden Hügel eingebettet und bietet sowohl einen Rückzugsort als auch einen Rahmen, um die Bewohner mit der Umgebung zu verbinden. Das Design bietet eine Reihe von miteinander verbundenen Erfahrungen mit pflegeleichten, verwitternden Materialien, die einen unfertigen Look hinterlassen, der sich in die Landschaft integriert. Offene Wohnbereiche verbinden sich mit Innenhöfen und Deckbereichen, die die Natur einfassen und kontrastierende Außenbereiche zum Sitzen und Essen bieten. Stampflehmwände sorgen für thermische Masse, und Hochleistungsglas, Belüftung und geothermische Konvektionssysteme sorgen für Komfort bei geringem Energieverbrauch. Das Haus dient als privater Rückzugsort für die Eigentümer und als Treffpunkt für die erweiterte Familie, mit einem Innenhofgarten in erhöhten Cortenbetten zur Produktion von Obst, Kräutern und Gemüse. Museumsqualität LED-Beleuchtung zeigt Kunst in ihrer besten Form und bietet essentielle Qualität ohne natürliches Licht.

Situé dans la vallée de Chittering, à une heure et demie au nord de Perth, le paysage est un mélange de zones de cultures défrichées et de forêts. La maison est nichée dans une colline en pente raide, créant à la fois un refuge et un cadre pour relier les occupants à l'environnement. Le design offre une série d'expériences interconnectées en utilisant des matériaux peu coûteux, résistant aux intempéries, avec une apparence inachevée qui s'intègre au paysage. Les espaces de vie ouverts se connectent avec des cours et des terrasses, encadrant la nature et offrant des espaces extérieurs contrastés pour s'asseoir et manger. Les murs en terre crue fournissent une masse thermique, et les vitrages performants, la ventilation et les systèmes de convection géothermiques maintiennent le confort avec une faible consommation d'énergie. La maison sert de refuge privé pour les propriétaires et de lieu de rencontre pour la famille élargie, avec un jardin intérieur dans des lits surélevés en corten pour produire des fruits, des herbes et des légumes. L'éclairage LED de qualité muséale permet d'afficher l'art dans toute sa splendeur et offre une qualité essentielle sans lumière naturelle.

Ubicada en el valle de Chittering, a una hora y media al norte de Perth, el paisaje es una mezcla de áreas de cultivo y bosques. La casa está ubicada en una ladera empinada, creando un refugio y un marco para conectar a los ocupantes con el entorno. El diseño ofrece una serie de experiencias interrelacionadas utilizando materiales de bajo mantenimiento y envejecidos, dejando un aspecto sin acabado que se integra con el paisaje. Los espacios de vida abiertos se conectan con patios y áreas de cubierta, enmarcando la naturaleza y proporcionando espacios al aire libre contrastantes para sentarse y comer. Las paredes de tierra apisonada proporcionan masa térmica, y los vidrios de alto rendimiento, la ventilación y los sistemas convectivos geotérmicos mantienen el confort con un bajo consumo de energía. La casa sirve como un retiro privado para los propietarios y un lugar de reunión para la familia extendida, con un jardín interior en camas elevadas de acero corten para producir frutas, hierbas y verduras. La iluminación LED de calidad de museo muestra el arte de la mejor manera posible y proporciona una calidad esencial sin luz natural.

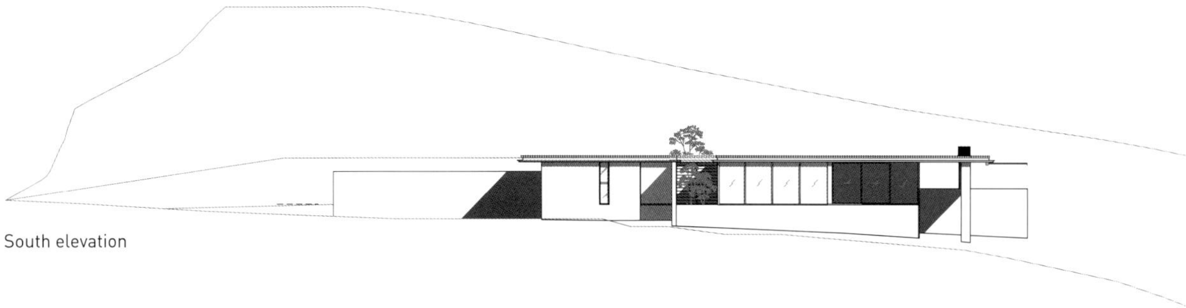

South elevation

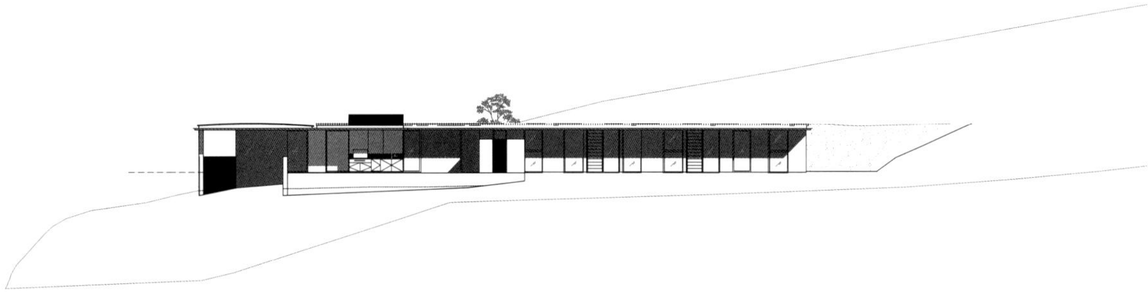

East elevation

Section A

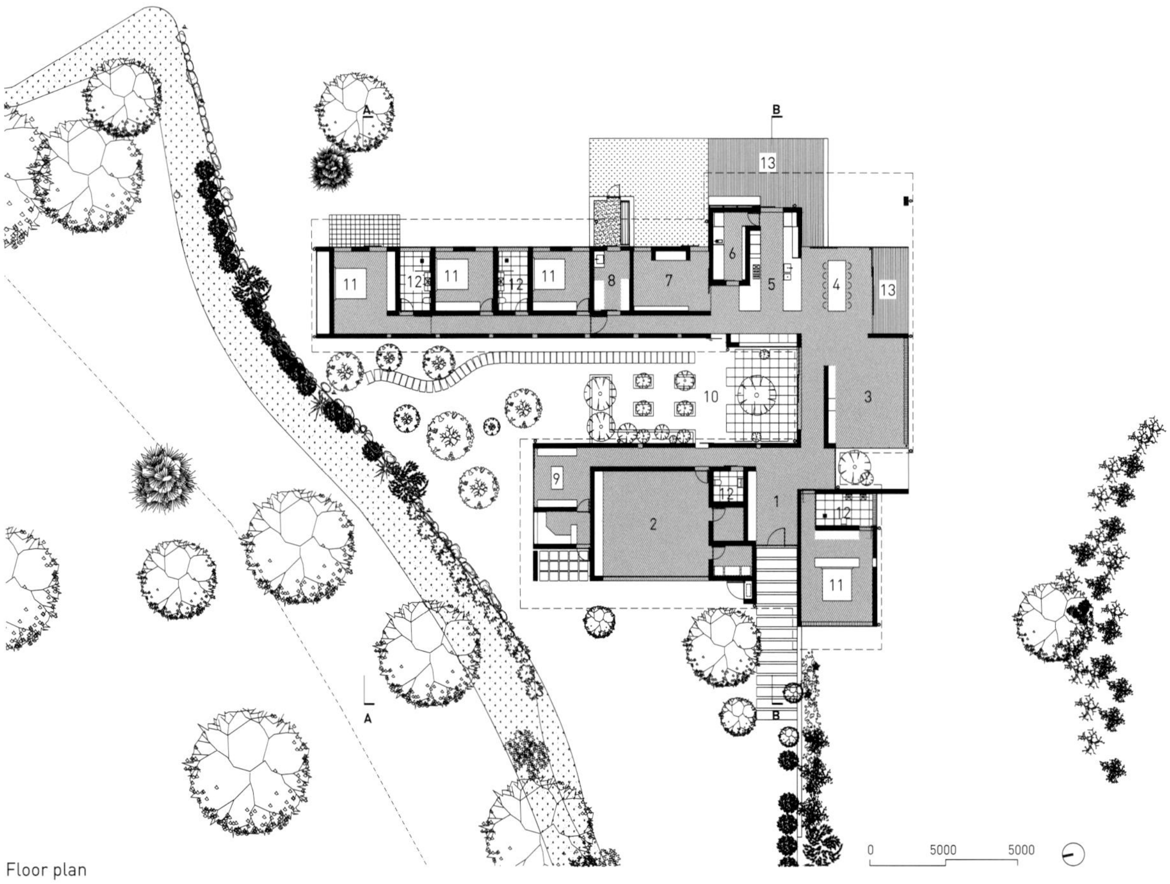

Floor plan

1. Entrance
2. Garage
3. Living room
4. Dining room
5. Kitchen
6. Pantry
7. Games room
8. Laundry
9. Office
10. Court
11. Room
12. Bathroom
13. Terrace

Ward | Blake Architects believes that architecture should respond to its particular place and be a natural partner with the environment. The firm's designs evolve from an appreciation for pragmatic solutions that show reverence to the bio-climatic environment in which they are placed. Since 1996, the firm has gained global recognition for architecture that successfully integrates with its surroundings and immerges the building users into a biophilic experience that blurs the lines between interior and exterior space.
By working within the context of each individual site and applying sustainable strategies Ward | Blake designs buildings that conserve energy and natural resources while creating high performance environments that bring well-being and delight to its occupants.

Ward | Blake Architects glaubt, dass Architektur auf ihren speziellen Ort reagieren und ein natürlicher Partner der Umwelt sein sollte. Die Designs der Firma entwickeln sich aus einer Wertschätzung für pragmatische Lösungen, die Ehrfurcht vor der bio-klimatischen Umgebung zeigen, in der sie platziert sind. Seit 1996 hat die Firma weltweite Anerkennung für Architektur gewonnen, die erfolgreich mit ihrer Umgebung integriert und die Gebäudenutzer in eine biophile Erfahrung eintaucht, die die Grenzen zwischen Innen- und Außenraum verschwimmen lässt. Durch Arbeit im Kontext jedes individuellen Standorts und Anwendung nachhaltiger Strategien entwirft Ward | Blake Gebäude, die Energie und natürliche Ressourcen einsparen und gleichzeitig Hochleistungsumgebungen schaffen, die Wohlbefinden und Freude für ihre Bewohner bringen.

TK PAD

Ward | Blake Architects estime que l'architecture devrait répondre à son lieu particulier et être un partenaire naturel de l'environnement. Les conceptions de l'entreprise évoluent à partir d'une appréciation des solutions pragmatiques qui montrent de la révérence pour l'environnement bio-climatique dans lequel elles se trouvent. Depuis 1996, l'entreprise a acquis une reconnaissance mondiale pour l'architecture qui s'intègre avec succès à son environnement et plonge les utilisateurs du bâtiment dans une expérience biophile qui brouille les lignes entre l'espace intérieur et extérieur. En travaillant dans le contexte de chaque site individuel et en appliquant des stratégies durables, Ward | Blake conçoit des bâtiments qui préservent l'énergie et les ressources naturelles tout en créant des environnements à haute performance qui apportent bien-être et plaisir à ses occupants.

Ward | Blake Architects cree que la arquitectura debe responder a su lugar particular y ser un compañero natural del medio ambiente. Los diseños de la firma evolucionan a partir de una apreciación por soluciones pragmáticas que muestran reverencia por el entorno bio-climático en el que se encuentran. Desde 1996, la firma ha ganado reconocimiento global por la arquitectura que se integra exitosamente con su entorno e inmerge a los usuarios del edificio en una experiencia biófila que difumina las líneas entre el espacio interior y exterior. Al trabajar dentro del contexto de cada sitio individual y aplicar estrategias sostenibles, Ward | Blake diseña edificios que conservan energía y recursos naturales al tiempo que crean ambientes de alto rendimiento que brindan bienestar y deleite a sus ocupantes.

WY NATURE CONSERVANCY

TK PAD

JACKSON, WYOMING, UNITED STATES // PHOTOS © PAUL WARCHOL

AWARDS:

- Award of Excellence
  Built Category AIA Wyoming
- 2009 Merit Award
  Custom Home Design Awards, Custom Detail Category
- 2008 Award of Excellence
  Wyoming Chapter of the American Institute of Architects

Design Architect: Thomas Ward, AIA
Client: Katherine Reedy
Client Project Managers: Stephen Kaness
Landscape Architect: N/A
Civil Engineer: Niskian Monks Engineers Ty Monks, PE

The site features wind-blown Loess fill from the Caldera of Yellowstone National Park and a unique outcrop of Conglomerate, remnants of a prehistoric inland sea. These geological features informed the design of the Residence, which is oriented along an east-west axis to face south. Wood-framed walls divide public and private spaces, and rammed earth walls utilize a patented post-tensioning steel system. Native soil and crusher fines are joined with Portland cement to form the walls, which offer passive thermal resistance and acoustic properties. The house envelope features elemental materials, including untreated concrete and native wood treated with water-borne sealers.

Die Website verfügt über windgetragenes Loess-Füllmaterial aus dem Caldera des Yellowstone-Nationalparks und einem einzigartigen Aufschluss von Konglomerat, Überresten eines prähistorischen Binnenmeeres. Diese geologischen Merkmale beeinflussten das Design der Residenz, die entlang einer Ost-West-Achse ausgerichtet ist, um nach Süden zu zeigen. Holzrahmenwände teilen öffentliche und private Räume, während gerammte Erdwände ein patentiertes Post-Spannungssystem aus Stahl nutzen. Einheimischer Boden und Brecherfeinmaterial werden mit Portlandzement verbunden, um die Wände zu bilden, die passive thermische Beständigkeit und akustische Eigenschaften bieten. Die Hausumhüllung besteht aus elementaren Materialien wie unbehandeltem Beton und einheimischem Holz, das mit wasserbasierten Dichtungsmitteln behandelt wird.

Le site présente une couche de loess transportée par le vent depuis la caldera du parc national de Yellowstone ainsi qu'une formation rocheuse unique de Conglomérat, vestige d'une mer intérieure préhistorique. Ces caractéristiques géologiques ont inspiré la conception de la résidence, qui est orientée selon un axe est-ouest pour faire face au sud. Des murs en bois divisent les espaces publics et privés, et des murs en terre compactée utilisent un système breveté de tension de l'acier. Le sol natif et les fines de concassage sont joints avec du ciment Portland pour former les murs, qui offrent une résistance thermique passive et des propriétés acoustiques. L'enveloppe de la maison est composée de matériaux élémentaires, notamment du béton non traité et du bois indigène traité avec des scellants à base d'eau.

El sitio cuenta con sedimentos de loess transportados por el viento desde la Caldera del Parque Nacional Yellowstone y una singular formación rocosa de conglomerado, restos de un antiguo mar interior. Estas características geológicas influyeron en el diseño de la residencia, que se orienta a lo largo de un eje este-oeste para enfrentar al sur. Las paredes con estructura de madera dividen los espacios públicos y privados, mientras que las paredes de tierra compactada utilizan un sistema patentado de tensión de acero. El suelo nativo y los residuos de trituración se unen con cemento Portland para formar las paredes, que ofrecen resistencia térmica pasiva y propiedades acústicas. El revestimiento de la casa utiliza materiales elementales, incluyendo hormigón sin tratar y madera autóctona tratada con selladores a base de agua.

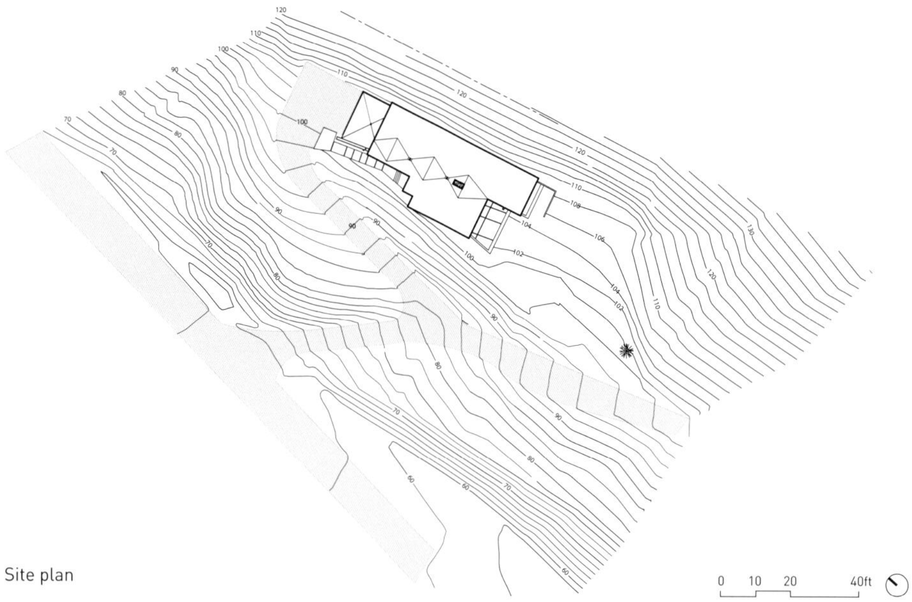

Site plan

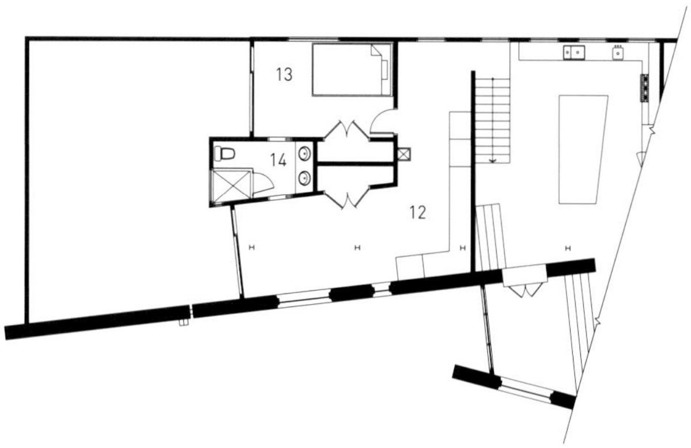

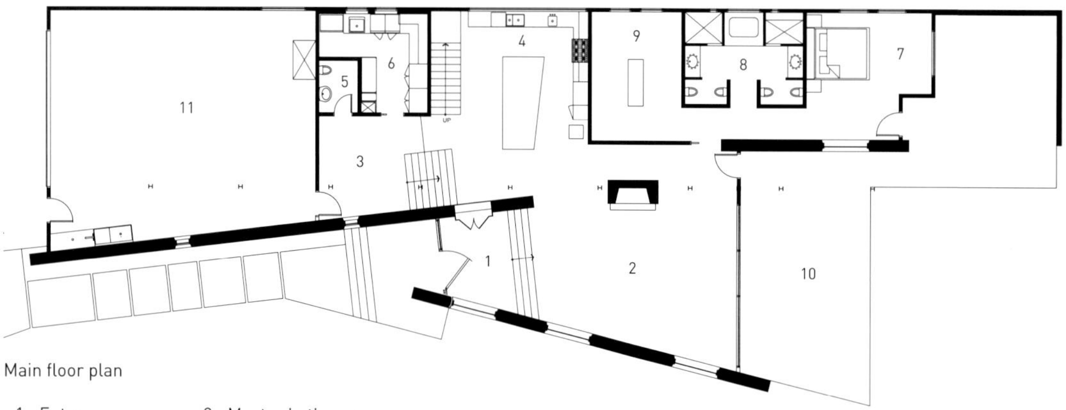

Main floor plan

1. Entry
2. Living
3. Mudroom
4. Kitchen
5. Powder room
6. Laundry
7. Master bedroom
8. Master bathroom
9. Master closet
10. Terrace
11. Garage
12. Office
13. Bedroom
14. Bathroom

0 5 10 20ft

WY NATURE CONSERVANCY

LANDER, WYOMING, UNITED STATES // PHOTOS © DOUGLAS KAHN

AWARDS:

- Award of Merit
  American Institute of Architects
- 2001 Merit Award
  Wyoming Chapter of the American Institute of Architects
- 1999 Honor Award
  Western Mountain Region/ American Institute of Architects

Design Architect: Thomas Ward, AIA
Client: Wyoming Chapter of The Nature Conservancy
Client Project Managers: Robert Carter, Ted Zimmerman
Civil Engineer: Joseph Grille, PE

The resource center at Red Canyon Ranch was designed using sustainable technology to capture a "sense of place" and demonstrate energy efficiency. The site, a gently sloping trailhead leading to the canyon, was chosen for its symbolic expression of the building as a "kiosk" directing visitors to areas of interest. Rammed earth was used as the primary building material to fulfill the conservancy's desire for low-impact technology, while the custom fabricated steel trusses provide a lightness contrasting the massiveness of the walls. Water conservation was critical, so a waterless composting toilet system and grey water disposal system were installed. The rammed earth walls provide thermal mass for even temperature levels, and lighting fixtures utilize compact fluorescent lamping and indirect halogen fixtures to reduce electric consumption and heat load.

Das Ressourcenzentrum am Red Canyon Ranch wurde unter Verwendung nachhaltiger Technologie entworfen, um ein „Ortsgefühl" zu vermitteln und die Energieeffizienz zu demonstrieren. Der Standort, ein sanft abfallender Trailhead, der zum Canyon führt, wurde aufgrund seines symbolischen Ausdrucks des Gebäudes als „Kiosk" gewählt, der Besucher zu interessanten Orten führt. Stampflehm wurde als Hauptbaumaterial verwendet, um dem Wunsch der Conservation nach einer Technologie mit geringem Einfluss zu entsprechen, während die maßgefertigten Stahlträger eine Leichtigkeit bieten, die im Gegensatz zur Massivität der Wände steht. Die Wassereinsparung war entscheidend, daher wurde ein wasserloses Komposttoiletten-System und ein Grauwasserentsorgungssystem installiert. Die Stampflehmwände bieten thermische Masse für gleichmäßige Temperaturen, und die Beleuchtung verwendet Kompaktleuchtstofflampen und indirekte Halogenleuch.

Le centre de ressources de Red Canyon Ranch a été conçu en utilisant une technologie durable pour capturer un « sens de l'endroit » et démontrer l'efficacité énergétique. Le site, une pente douce menant au canyon, a été choisi pour son expression symbolique du bâtiment comme un « kiosque » dirigeant les visiteurs vers des zones d'intérêt. La terre battue a été utilisée comme matériau de construction principal pour répondre au désir de la conservation d'une technologie à faible impact, tandis que les fermes en acier fabriquées sur mesure fournissent une légèreté qui contraste avec la massivité des murs. La conservation de l'eau était critique, donc un système de toilettes compostables sans eau et un système d'élimination des eaux grises ont été installés. Les murs en terre battue fournissent une masse thermique pour des niveaux de température uniformes, et les luminaires utilisent des lampes fluorescentes compactes et des luminaires halogènes indirects pour réduire la consommation d'électricité et la charge thermique.

El centro de recursos en Red Canyon Ranch fue diseñado utilizando tecnología sostenible para capturar un «sentido de lugar» y demostrar eficiencia energética. El sitio, un sendero suavemente inclinado que conduce al cañón, fue elegido por su expresión simbólica del edificio como un «kiosco» que dirige a los visitantes a áreas de interés. Se utilizó tierra apisonada como material de construcción principal para cumplir con el deseo de la conservación de tecnología de bajo impacto, mientras que las cerchas de acero fabricadas a medida proporcionan una ligereza que contrasta con la masividad de las paredes. La conservación del agua fue crítica, por lo que se instaló un sistema de inodoros composteros sin agua y un sistema de eliminación de aguas grises. Las paredes de tierra apisonada proporcionan masa térmica para niveles de temperatura uniformes, y las luminarias utilizan lámparas fluorescentes compactas y luminarias halógenas indirectas para reducir el consumo de energía eléctrica y la carga térmica.

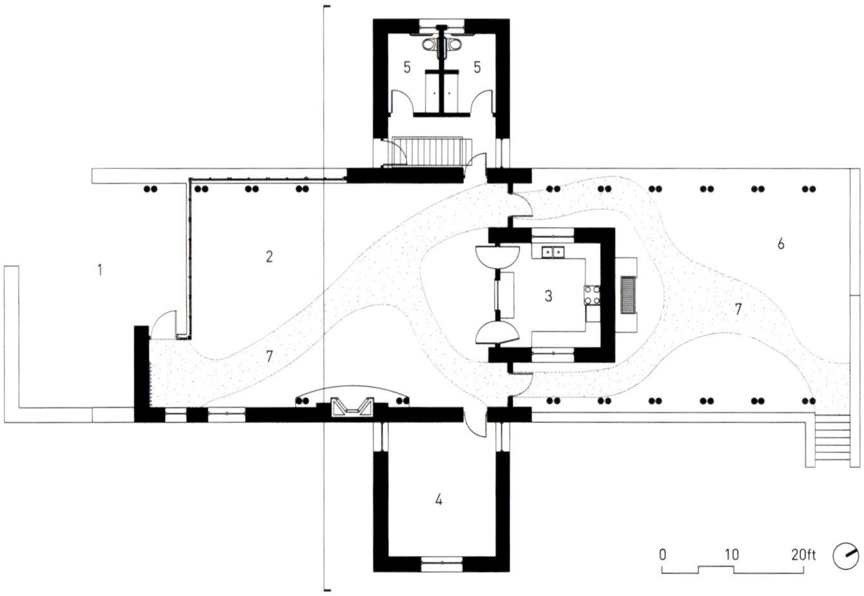

1. Entry patio
2. Meeting room
3. Kitchen
4. Office
5. Bathroom
6. Outdoor patio
7. Trail

Main floor plan

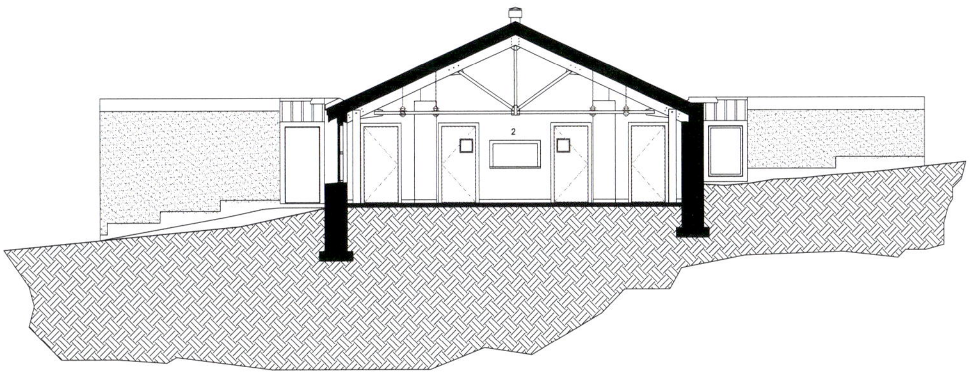

Section

ZAV Architects aims to streamline the architectural production process by reimagining components such as diagrams, scenarios, crafts, and material pallets based on daily life. The approach begins with observing everyday realities and analyzing data to identify opportunities for alteration. Playfulness is key to adapting and experimenting with content changes in an adaptive manner. This alternative approach to architecture is achieved by allowing for modification and transformation, as demonstrated in projects such as "Majara" and "Type-less." "Orphan Habitat" redefines interior and exterior spaces through dynamic scaffold curtains, while "Farsh Film" embraces the imperfections of every craft. "Saraye Pedari" offers a contemporary take on an existing building influenced by accessible crafts, and the "Well-Being Center" transforms the traditional nomadic tent into a colorful canopy. This "playful relevant" architecture prioritizes popular materials and methods, resulting in a welcoming and beautiful end product.

ZAV Architects zielt darauf ab, den architektonischen Produktionsprozess zu optimieren, indem sie Komponenten wie Diagramme, Szenarien, Handwerkskunst und Materialpaletten basierend auf dem täglichen Leben neu erfindet. Der Ansatz beginnt mit der Beobachtung des Alltags und der Analyse von Daten, um Möglichkeiten für Veränderungen zu identifizieren. Spielfreude ist der Schlüssel zur Anpassung und Experimentierung mit Inhaltänderungen auf adaptive Weise. Dieser alternative Ansatz zur Architektur wird durch die Möglichkeit von Modifikation und Transformation erreicht, wie es bei Projekten wie „Majara" und „Type-less" demonstriert wird. „Orphan Habitat" definiert Innen- und Außenräume durch dynamische Gerüstvorhänge neu, während „Farsh Film" die Unvollkommenheiten jedes Handwerks umarmt. „Saraye Pedari" bietet eine zeitgenössische Interpretation eines bestehenden Gebäudes, beeinflusst durch zugängliche Handwerkskunst, und das „Well-Being Center" verwandelt das traditionelle Nomadenzelt in ein buntes Zelt. Diese „spielerische relevante" Architektur priorisiert beliebte Materialien und Methoden, was zu einem einladenden und schönen Endprodukt führt.

PRESENCE IN HORMUZ 2

ZAV Architects vise à simplifier le processus de production architecturale en réinventant des composants tels que les diagrammes, les scénarios, l'artisanat et les palettes de matériaux en fonction de la vie quotidienne. L'approche commence par l'observation des réalités quotidiennes et l'analyse des données pour identifier les opportunités de changement. La ludicité est clé pour s'adapter et expérimenter avec les changements de contenu de manière adaptative. Cette approche alternative de l'architecture est réalisée en permettant la modification et la transformation, comme le démontrent des projets tels que « Majara » et « Type-less ». « Orphan Habitat » redéfinit les espaces intérieurs et extérieurs grâce à des rideaux d'échafaudage dynamiques, tandis que « Farsh Film » embrasse les imperfections de chaque métier. « Saraye Pedari » offre une vision contemporaine d'un bâtiment existant influencé par l'artisanat accessible, et le « Well-Being Center » transforme la tente nomade traditionnelle en un auvent coloré. Cette architecture « ludique pertinente » priorise les matériaux et les méthodes populaires, ce qui se traduit par un produit final accueillant et beau.

ZAV Architects tiene como objetivo simplificar el proceso de producción arquitectónica al reinventar componentes como diagramas, escenarios, artesanías y paletas de materiales basados en la vida diaria. El enfoque comienza con la observación de las realidades cotidianas y el análisis de datos para identificar oportunidades de cambio. La juguetona adaptación y experimentación con los cambios de contenido es clave para este enfoque alternativo de la arquitectura, el cual se logra mediante la modificación y transformación, como se evidencia en proyectos como «Majara» y «Type-less». «Orphan Habitat» redefine los espacios interiores y exteriores a través de dinámicas cortinas de andamiaje, mientras que «Farsh Film» abraza las imperfecciones de cada artesanía. «Saraye Pedari» ofrece una visión contemporánea de un edificio existente influenciado por artesanías accesibles, y el «Well-Being Center» transforma la tradicional tienda nómada en un colorido dosel. Esta arquitectura «juguetona relevante» prioriza los materiales y métodos populares, resultando en un producto final acogedor y hermoso.

PRESENCE IN HORMUZ 2

HORMUZ ISLAND, IRAN
PHOTOS © TAHMINEH MONZAVI, SOROUSH MAJIDI, PAYMAN BARKHORDARI

Architecture Firm: ZAV Architects
Completion Year: 2020
Gross Built Area: 10,300 m²
Area: 4,000 m²
Landscape Area: 6,300 m²

Lead Architects: Mohamadreza Ghodousi, Fatemeh Rezaei, Golnaz Bahrami, Soroush Majidi
Design Assistants: Sheila Ehsaei,Sara Jafari, Payman Barkhordari, Mohsen Safshekan, Kaveh Rashidzadeh, Hossein Panjehpour
Landscape: Maryam Yousefi, Morteza Adib
Interior Design: Sara Jafari, Taraneh Behboud, Sara Nikkar, Mohsen Dehghan
Light Consultant: Tajang Light
Supervision: Soroush Majidi, Payman Barkhordari, Sheila Ehsaei
Modeling: Somayeh Saeidi
Presentation: Fereshteh Assadzadeh, Somayeh Saeidi, Arshia Hashemipour, Dorsa Tavakoli, Sara Fallahzadeh

Owner: Ali Rezvani
Client: Ehsan Rasoulof

Structural Design: Behrang Baniadam, Rouhi Touski
Civil Engineering: Farhad Beigi
Electrical Engineering: Pejman Moradian
Mechanical Engineering: Saeid Afsharian
Project Constructor: Amir Tehrani Nobahari
Construction Manager: Hormat Ghasemi
Mechanical Constructor: Javad Irandegani, Hamid Haji Posht-e-Gol
Floor Constructor: Davoud Etemadi
Fenestration Builder: Mehra Company
Interior Plaster: Gholamali Abbasi

The Presence in Hormuz project in Iran aims to expand the impact of architecture beyond the built environment by activating certain points on the island of Hormuz. Majara Residence is one of the projects, which consists of 200 organic earth domes that offer accommodations for tourists and public usages open to all. The project's architecture is informed by the island's colorful landscapes and porous green fabric consisting of native species of trees. The use of a polychromic pallet is tuned with the colorful context, and the formal connection of the architectural proposal is further enhanced through the color pallet. The project is established out of a variegated taxonomy of molecular structures, creating a constructed replica of the geographical skyline, and it was built using a low-tech method called superadobe, empowering local workers in the process. The result is a polychromic gateless neighborhood containing public activities built by the hands of the Hormuzi people.

Das Presence in Hormuz Projekt im Iran zielt darauf ab, den Einfluss der Architektur über die gebaute Umgebung hinaus zu erweitern, indem bestimmte Punkte auf der Insel Hormuz aktiviert werden. Die Majara Residence ist eines der Projekte, das aus 200 organischen Erdkuppeln besteht, die Unterkünfte für Touristen und öffentliche Nutzungen für alle bieten. Die Architektur des Projekts orientiert sich an den farbenfrohen Landschaften der Insel und dem porösen grünen Gewebe, das aus einheimischen Baumarten besteht. Die Verwendung eines polychromen Farbschemas stimmt mit dem farbenfrohen Kontext überein, und die formale Verbindung des architektonischen Vorschlags wird durch das Farbschema weiter verbessert. Das Projekt basiert auf einer vielfältigen Taxonomie von Molekülstrukturen und schafft eine konstruierte Nachbildung des geografischen Horizonts. Es wurde mit einer Low-Tech-Methode namens Superadobe gebaut, wodurch lokale Arbeiter gestärkt werden. Das Ergebnis ist eine polychrome, torlose Nachbarschaft mit öffentlichen Aktivitäten, die von den Händen der Hormuzi-Bevölkerung gebaut wurde.

Le projet Presence in Hormuz en Iran vise à étendre l'impact de l'architecture au-delà de l'environnement construit en activant certains points de l'île de Hormuz. La résidence Majara est l'un des projets, qui se compose de 200 dômes organiques en terre offrant des hébergements pour les touristes et des espaces publics ouverts à tous. L'architecture du projet est informée par les paysages colorés de l'île et le tissu vert poreux constitué d'espèces d'arbres indigènes. L'utilisation d'une palette polychrome est accordée avec le contexte coloré, et la connexion formelle de la proposition architecturale est encore renforcée par la palette de couleurs. Le projet est établi à partir d'une taxonomie variée de structures moléculaires, créant une réplique construite de l'horizon géographique, et il a été construit en utilisant une méthode à faible technologie appelée superadobe, renforçant ainsi les travailleurs locaux. Le résultat est un quartier polychrome sans portes contenant des activités publiques construites par les mains du peuple Hormuzi.

El proyecto Presence en Hormuz en Irán tiene como objetivo expandir el impacto de la arquitectura más allá del entorno construido, activando ciertos puntos en la isla de Hormuz. Majara Residence es uno de los proyectos, que consiste en 200 cúpulas orgánicas de tierra que ofrecen alojamiento para turistas y uso público abierto a todos. La arquitectura del proyecto está informada por los paisajes coloridos de la isla y la porosa tela verde compuesta por especies nativas de árboles. El uso de una paleta policromática se sintoniza con el contexto colorido, y la conexión formal de la propuesta arquitectónica se mejora aún más a través de la paleta de colores. El proyecto se establece a partir de una taxonomía variada de estructuras moleculares, creando una réplica construida del horizonte geográfico, y fue construido utilizando un método de baja tecnología llamado superadobe, empoderando a los trabajadores locales en el proceso. El resultado es un vecindario policromático sin puertas que contiene actividades públicas construidas por las manos de la gente de Hormuz.

Sections

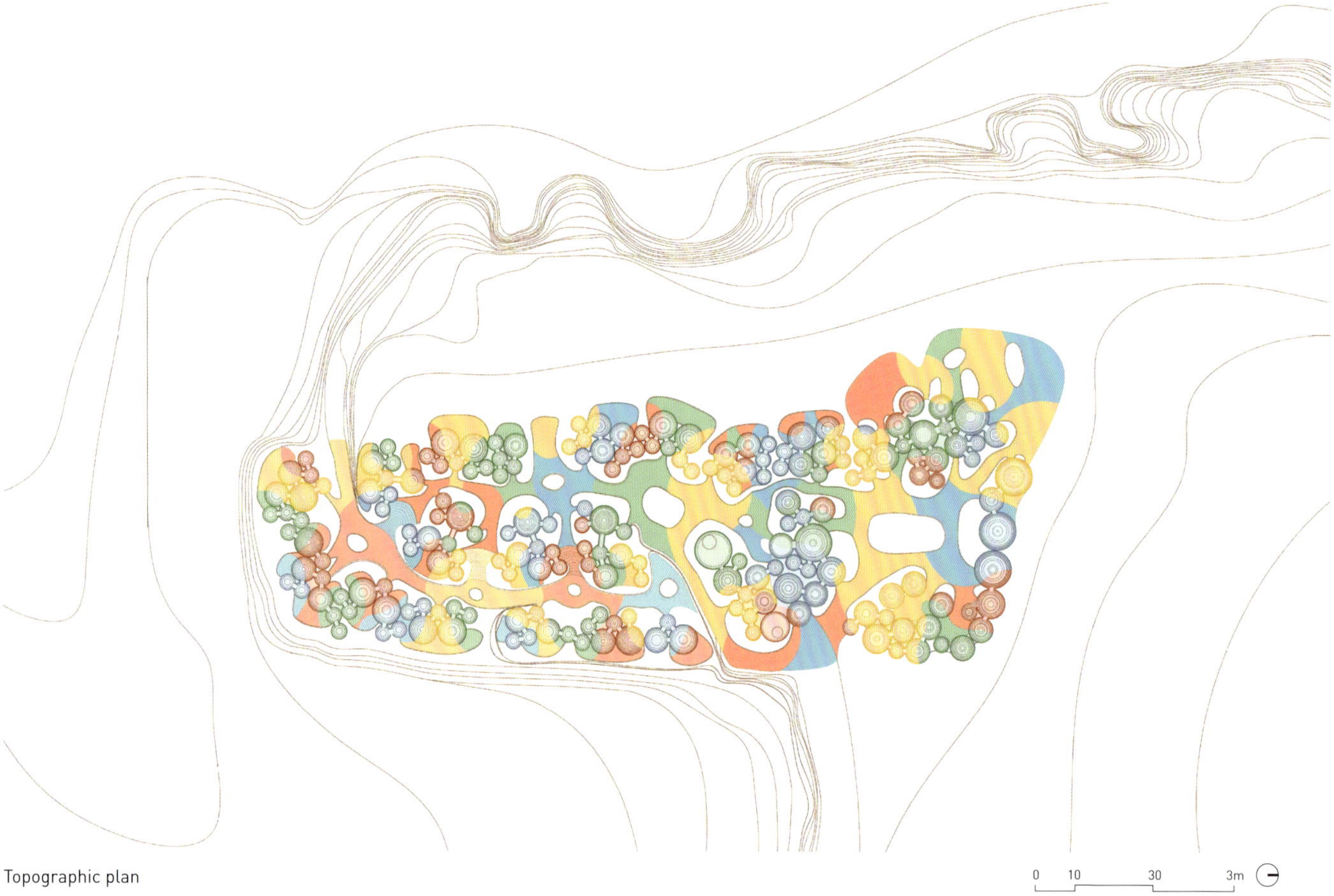

Topographic plan

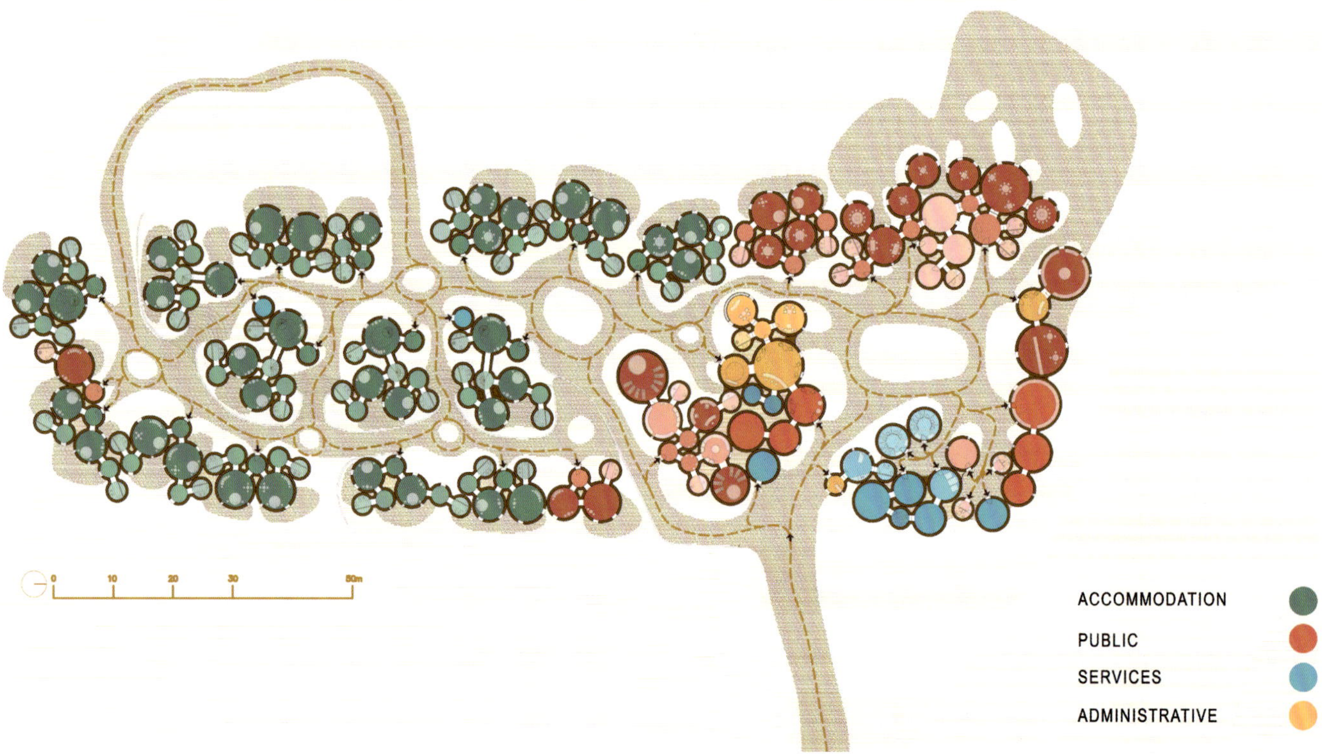

Physical program